Pretreatment of Industrial Wastes

Manual of Practice No. FD-3

Prepared by the **Task Force on Pretreatment of Industrial Wastes**
Elin Eysenbach, *Chair*

Michael R. Alberi	Byung R. Kim	Steven A. Shedroff
Paul R. Anderson	Jeffrey L. Lape	Sam Shelby
W. Bailey Barton	Charles D. Malloch	Dennis P. Shelly
Peter V. Cavagnaro	Michael P. McGinness	Rao Y. Surampalli
Harry Criswell	William J. Mikula	William M. Throop
Charles Darnell	Stephen J. Miller	Paul J. Usinowicz
Michael C. Downey	Larry W. Moore	Gary R. Vaughan
Terence P. Driscoll	John L. Musterman	Kannan Vembu
Richard C. Grant	Tom M. Pankratz	T. Viraraghavan
Negib Harfouche	David M. Philbrook	Thomas G. Wallace
Roger R. Hlavek	Douglas L. Ralston	Hugh E. Wise, Jr.
W. Michael Joyce	Herbert N. Schott	George M. Wong-Chong
Robert N. Kenney	Blaine F. Severin	

Under the Direction of the
Facilities Development Subcommittee of the Technical Practice Committee
and sponsored by the
Industrial Wastes Committee

1994

Water Environment Federation
601 Wythe Street
Alexandria, Virginia 22314–1994 USA

The Water Environment Federation is a nonprofit, educational organization composed of member and affiliated associations throughout the world. Since 1928, WEF has represented water quality specialists, including biologists, bacteriologists, local and national government officials, treatment plant operators, laboratory technicians, chemists, industrial technologists, students, academics, equipment manufacturers/distributors, and civil, design, and environmental engineers.

For information on membership, publications, and conferences, contact
    Water Environment Federation
    601 Wythe Street
    Alexandria, VA 22314–1994 USA
    (703) 684-2400

Library of Congress Cataloguing-in-Publication Data
Pretreatment of industrial wastes / prepared by the Task Force on Pretreatment of Industrial Wastes; under the direction of the Facilities Development Subcommittee of the Technical Practice Committee and sponsored by the Industrial Wastes Committee.
        p.   cm.   —(Manual of practice; no. FD–3)
    Includes bibliographical references and index.
    ISBN 1-881369-89-7: $55.00
        1. Factory and trade waste—Purification.   2. Sewage—Purification.
I. Water Environment Federation. Task Force on Pretreatment of Industrial Wastes.   II. Water Environment Federation. Facilities Development Subcommittee. III. Water Environment Federation. Industrial Wastes Committee.
IV. Series: Manual of practice.   FD; no. 3.
TD897.5.P75   1994   94-30220
628.3'4—dc20   CIP

Manuals of Practice for Water Pollution Control

The Water Environment Federation Technical Practice Committee (formerly the Committee on Sewage and Industrial Wastes Practice of the Federation of Sewage and Industrial Wastes Associations) was created by the Federation Board of Control on October 11, 1941. The primary function of the Committee is to originate and produce, through appropriate subcommittees, special publications dealing with technical aspects of the broad interests of the Federation. These manuals are intended to provide background information through a review of technical practices and detailed procedures that research and experience have shown to be functional and practical.

IMPORTANT NOTICE

The contents of this publication are for general information only and are not intended to be a standard of the Water Environment Federation (WEF).

No reference made in this publication to any specific method, product, process, or service constitutes or implies an endorsement, recommendation, or warranty thereof by WEF.

WEF makes no representation or warranty of any kind whether expressed or implied, concerning the accuracy, product, or process discussed in this publication and assumes no liability.

Anyone utilizing this information assumes all liability arising from such use, including but not limited to infringement of any patent or patents.

Water Environment Federation Technical Practice Committee Control Group

F.D. Munsey, *Chair*
L.J. Glueckstein, *Vice-Chair*

C.N. Lowery
R.W. Okey
T. Popowchak
J. Semon

Authorized for Publication by the Board of Control
Water Environment Federation

Quincalee Brown, *Executive Director*

Preface

This manual is a joint effort by the Federation's Technical Practice and Industrial Wastes Committees to provide general guidance on treatment selection, design, facilities development, and pollution prevention for the pretreatment of industrial wastes. The manual is targeted for industrial users of wastewater treatment plants (WWTP) (indirect dischargers), WWTP officials implementing pretreatment programs, and the technical people who support pretreatment design and development. Industrial indirect dischargers having the most to gain from this manual are those who need to pretreat wastewater for compliance with pretreatment ordinances. As an alternative to pretreatment itself, the manual also discusses in-plant pollution prevention techniques.

The manual begins with an overview of pretreatment regulatory requirements, potential management strategies, and pollution prevention considerations. The remainder of the manual is organized by waste characteristics. Chapters address the need for pretreatment of the specific waste characteristic and the technology selection and design considerations important in treating that characteristic.

This manual was produced under the direction of Elin Eysenbach, *Chair*. The principal authors are

Terence P. Driscoll	Douglas L. Ralston
Elin Eysenbach	Herbert N. Schott
W. Michael Joyce	Steven A. Shedroff
Charles D. Malloch	Gary R. Vaughan
John L. Musterman	George M. Wong-Chong

Contributing authors are

Michael R. Alberi	Thomas M. Pankratz
Don Bzdyl	David N. Philbrook
Peter V. Cavagnaro	Stephen R. Tate
Richard C. Grant	Steve Tripmacher
Charles C. Meyer	Hugh E. Wise, Jr.
Larry W. Moore	

Additional information and review were provided by

Paul R. Anderson	Michael P. McGinness
W. Bailey Barton	William J. Mikula
Charles Darnell	Stephen J. Miller
Jim Devlin	Blaine Severin
Michael C. Downey	Sam Shelby
John Groenewold	Dennis P. Shelly
Samuel J. Hadeed	Rao Y. Surampalli
Negib Harfouche	William M. Throop
Roger R. Hlavek	Paul J. Usinowicz
Robert N. Kenney	Kannan Vembu
Byung R. Kim	T. Viraraghavan
Jeffrey L. Lape	Thomas G. Wallace

Task force members', reviewers', and authors' efforts were supported by the following organizations:

Advent Group Inc., Brentwood, Tennessee
Alden Environmental Management, Inc., Wayne, Pennsylvania
Aponowich, Driscoll & Associates, Inc., Atlanta, Georgia
Aqua-Chem, Inc., Houston, Texas
Arthur D. Little, Inc., Cambridge, Massachusetts
Bio-Cide International, Inc., Lincoln, Nebraska
Borden Environmental Affairs, Columbus, Ohio
Calderon-Grant, Inc., Columbus, Ohio
City of Austin, Austin, Texas
Consulting Analytical Services, Springfield, Missouri
Dufresne-Henry, North Springfield, Vermont
Eckenfelder Inc., Nashville, Tennessee
Engineering-Science, Liverpool, New York
ENSR Consulting & Engineering, Somerset, New Jersey
ERM, Inc., Exton, Pennsylvania
ICF-Kaiser Engineers, Pittsburgh, Pennsylvania
Illinois Institute of Technology, Chicago, Illinois
John Carollo Engineers, West Linn, Oregon
Jordan Jones & Goulding, Snellville, Georgia
McNamee Operational Services, Ann Arbor, Michigan
Monsanto Company, St. Louis, Missouri
Nolte & Associates, Sacramento, California
OMI, Inc., Fayetteville, Arkansas
Procter & Gamble, Cincinnati, Ohio
Proctor Davis & Ray Engineers, Lexington, Kentucky
R.E. Wright Associates, Inc., Mechanicsburg, Pennsylvania

Roy F. Weston, Inc., Edison, New Jersey
RUST E & I, Raleigh, North Carolina
Sadat Associates, Inc., Princeton, New Jersey
Seeler Associates, Rochester, New York
Stanley Consultants, Muscatine, Iowa
Sverdrup Corporation, Maryland Heights, Missouri
U.S. Environmental Protection Agency, Washington, D.C.
Unified Sewerage Agency, Hillsboro, Oregon
Union Sanitary District, Fremont, California
University of Houston, Pasadena, Texas
University of Regina, Regina, Saskatchewan, Canada
Woodard & Curran, Inc., Portland, Maine

Federation technical staff project management was provided by Eileen J.
O'Neill and J. Robert Schweinfurth. Technical editorial assistance was provided by Matthew Hauber.

Contents

List of Tables

List of Figures

Chapter 1
Introduction

PURPOSE AND SCOPE OF MANUAL

The objective of this manual is to provide guidance in the selection of designs and processes for the pretreatment of industrial wastes. It is directed toward industrial users of municipal wastewater treatment plants (WWTP) (indirect dischargers), WWTP officials implementing pretreatment programs, and technical personnel who support pretreatment design and development. Industrial indirect dischargers who should benefit most from this manual are those who must pretreat wastewater for compliance with pretreatment ordinances. Although this manual includes general discussions of design and operations, design details and operational specifics are beyond its scope. As an alternative to meet pretreatment objectives, the manual also discusses in-plant pollution prevention techniques.

The first three chapters set the stage for pretreatment process selection by discussing the background and issues applicable to pretreatment in general. The first chapter demonstrates the need for pretreatment information and guidance and provides a framework for later waste-specific discussions.

Although the manual is not a legal or regulatory advisory, a basic understanding of the U.S. regulatory status regarding pretreatment, as discussed in Chapter 2, sets the stage for understanding the need for pretreatment and the driving forces for technology selection. Those seeking situational clarification should refer to relevant regulations.

Chapter 3 addresses wastewater management alternatives and pollution prevention (in-plant controls). The remaining chapters offer specific selection and implementation guidance for meeting pretreatment requirements. They are organized by waste characteristics, as follows:

Chapter 4: Flow Equalization
Chapter 5: Solids Separation and Handling
Chapter 6: Fats, Oil, and Grease Removal
Chapter 7: Neutralization
Chapter 8: Heavy Metals Removal
Chapter 9: Treatment of Organic Constituents
Chapter 10: Nutrient Removal

Allowing for some overlap, Chapters 4 to 6 generally address physical processes; Chapters 7 and 8, chemical; and Chapters 9 and 10, biological.

Although wastewater conveyed to a WWTP may include process wastewater from industrial users, in-sewer dilution by domestic wastewater typically will substantially reduce the original concentrations of chemicals in the industrial discharges. Overall treatment efficiency, however, may be temporarily diminished by higher concentrations (slugs) of chemicals to which the biomass in the WWTP's treatment process have not previously adapted. Temporary operational problems may also occur when slugs enter the sewer too close to the WWTP for adequate dilution or during periods of the day when the flow of domestic wastewater is low.

For the purposes of this manual, the U.S. regulatory definition will be used for pretreatment. It is "the reduction of the amount of pollutants, the elimination of pollutants, or the alteration of the nature of pollutant properties in wastewater prior to, or in lieu of, discharging or otherwise introducing such pollutants into a waste treatment plant."

A manufacturing facility discharging to a WWTP may have to pretreat water to

- Comply with the General Pretreatment Regulations issued by the U.S. Environmental Protection Agency (U.S. EPA);
- Comply with Categorical Pretreatment Standards, if applicable;
- Comply with local ordinances (limits);
- Reduce user fees, when the surcharge (by the WWTP) is based in part on the mass loading of one or more pollutants; and
- Improve its public image or reduce the stigma associated with publicly reported releases such as the "Toxics Release Inventory" under the Superfund Amendment and Reauthorization Act (SARA).

The emphasis of this manual is on indirect discharges with final treatment done by the WWTP in conjunction with the treatment of residential sanitary wastes. Most of the technologies discussed are also applicable to industrial wastes treated within the industrial facility (direct discharger) or by an off-site, privately owned wastewater treatment facility.

Specific wastewater management alternatives and strategies for pretreatment are discussed in Chapter 3. Although they are not new to industry, emphasis has been placed recently on pollution prevention techniques. These techniques are in-plant process designs or redesigns that avoid or reduce wastewater discharge before and, to the extent possible, instead of effluent treatment. Although not pretreatment as such, pollution prevention is critical to a sound wastewater management strategy.

For the many affected industrial facilities to meet these requirements, substantial investments in the construction and operation of pretreatment facilities has been, and will continue to be, necessary. This manual provides general treatment selection and design guidance, and facilities development information for the pretreatment of industrial wastewater. The reader should refer to detailed wastewater design texts for additional information. Several of these are included in the reference lists provided by chapter.

The reader is cautioned against applying general values taken from this manual to specific industrial applications. To ensure proper design, site-specific evaluations and appropriate treatability testing will be needed.

INDUSTRIAL WASTEWATER CHARACTERISTICS

Industrial processes generate a variety of wastewater pollutants, some of which are difficult and costly to treat. Wastewater characteristics and levels of pollutants vary significantly from one industry to another and sometimes vary greatly among plants within the same industry. In addition, fluctuations may occur in the wastewater's daily and hourly flow, temperature, and composition. In some cases, several manufacturing processes may discharge at the same location. Together these factors create a complex wastewater management challenge that is unique to each industrial facility.

Pollutants regulated by U.S. pretreatment standards may be grouped into conventional, nonconventional, and priority pollutants, as shown in Table 1.1. Tables 1.2 to 1.6 divide priority pollutants according to waste characteristics. Knowledge of potential wastewater components and awareness of these regulated discharges will allow assessment of the applicability of pretreatment standards to a given facility. A list of pretreatment technologies commonly applied to pollutants regulated by pretreatment standards is shown in Table 1.7. Chapters 4 through 10 will address the applications of such technologies.

Table 1.1   Pollutants regulated by pretreatment standards.

Priority pollutants	Nonconventional pollutants
Asbestos (as TSS)	Ammonia (as nitrogen)
Cyanide, total	Chromium VI (hexavalent)
Metals, total	Chemical oxygen demand (COD)
Organics	COD/BOD$_7$ (7-day)
Volatile	Fluoride
Semivolatile	Manganese
Total toxic organics (TTO)*	Nitrate (as N)
Conventional pollutants	Organic nitrogen (as N)
Biochemical oxygen	Pesticide active ingredients (PAI)
demand, 5-day (BOD$_5$)	Phenols, total
Total suspended solids (TSS)	Phosphorus, total (as P)
Oil and grease	Total organic carbon (TOC)
Oil (animal and vegetable)	
Oil (mineral)	
pH	

*   TTO is not always defined as the same organic priority pollutants (see 40 CFR Parts 413.02, 464.02, 467.02, and 469.12).

Table 1.2   Priority pollutants—metals (information taken from 40 CFR Part 423, Appendix A).

Antimony, total	Chromium, total	Nickel, total	Zinc, total
Arsenic, total	Copper, total	Selenium, total	
Beryllium, total	Lead, total	Silver, total	
Cadmium, total	Mercury, total	Thallium, total	

Table 1.3   Priority pollutants—volatile organics.

Acrolein	1,2-*trans*-Dichloroethylene
Acrylonitrile	1,2-Dichloropropane
Benzene	1,3-Dichloropropylene
Bromodichloromethane	Ethylbenzene
Bromoform	Methyl bromide
Carbon tetrachloride	Methyl chloride
Chlorobenzene	Methylene chloride
Chlorodibromomethane	1,1,2,2,-Tetrachloroethane
Chloroethane	Tetrachloroethylene
2-Chloroethylvinyl ether	Toluene
Chloroform	1,1,1-Trichloroethane
1,1-Dichloroethane	1,1,2-Trichloroethane
1,2-Dichloroethane	Trichloroethylene
1,1-Dichloroethylene	Vinyl chloride

Table 1.4   Priority pollutants—semivolatile organics.

Acenaphthene	2,4-Dinitrotoluene
Acenaphthylene	Di-n-octyl phthalate
Antracene	1,2-Diphenylhydrazine (as azobenzene)
Benzidine	Fluoranthene
Benzo(a)anthracene	Fluorene
Benzo(b)fluoranthene	Hexachlorobenzene
Benzo(a)pyrene	Hexachlorobutadiene
Benzo(ghi)perylene	Hexachlorocyclopentadiene
4-Bromophenylphenyl ether	Hexachloroethane
Butylbenzyl phthalate	Indeno(1,2,3-cd)pyrene
4-Chloro-m-cresol	Isophorone
Bis(2-Chloroethoxy)methane	Naphthalene
Bis(2-Chloroisopropyl) ether	Nitrobenzene
Chlorophenol	2-Nitrophenol
2-Chloronaphthalene	4-Nitrophenol
Chrysene	N-Nitrosodimethylamine
Dibenzo(a,h)anthracene	N-Nitrosodiphenylamine
1,2-Dichlorobenzene	N-Nitrosodipropylamine
1,3-Dichlorobenzene	Pentachlorophenol
1,4-Dichlorobenzene	Phenanthrene
3,3-Dichlorobenzidine	Phenol
2,4-Dichlorophenol	Polychlorinated biphenyls (PCBs)
Dibutyl phthalate	PCB-1016 PCB-1248
Diethyl phthalate	PCB-1221 PCB-1254
Di(2-ethylhexyl)phthalate	PCB-1232 PCB-1260
Dimethyl phthalate	PCB-1242
2,4-Dimethylphenol	Pyrene
2,4-Dinitrophenol	2,3,7,8-Tetrachlorodibenzo-p-dioxin (TCDD)
4,6-Dinitro-o-cresol	1,2,4-Trichlorobenzene
2,6-Dinitrotoluene	

Table 1.5   Priority pollutants—pesticides (information taken from 40 CFR Part 423, Appendix A).

Aldrin	4,4′-DDT	Endosulfan sulfate
α-BHC	4,4′-DDD	Endrin
β-BHC	4,4′-DDE	Endrin aldehyde
δ-BHC	Dieldrin	Heptachlor
γ-BHC	α-Endosulfan	Heptachlor epoxide
Chlordane	β-Endosulfan	Toxaphene

Table 1.6   Pesticide active ingredients regulated by pretreatment standards (FR 57 at 12598 [April 10, 1992] and 40 CFR Part 455.20[d]).

Acephate	Diazinon	Naled
Acifluorfen	Dichlorprop	Norflurazon
Alachlor	Dichlorvos	Organo-tin
Aldicarb	Dinoseb	Parathion
Ametryn	Dioxathion	Parathion methyl
Atrazine	Disulfoton	PCNB
Azinphos methyl	Diuron	Pendimethalin
Benfluralin	Endothall	Permethrin
Benomyl	Endrin*	Phorate
Biphenyl	Ethalfluralin	Phosmet
Bolstar	Ethion	Prometon
Bromacil	Fenarimol	Prometryn
Bromacil lithium	Fensulfothion	Pronamide
Bromoxynil	Fenthion	Propachlor
Bromoxynil octanoate	Fenvalerate	Propanil
Busan 40	Glyphosate	Propazine
Busan 85	Heptachlor*	Pyrethrin I
Butachlor	Isopropalin	Pyrethrin II
Captafol	KN methyl	Simazine
Carbam-*S*	Linuron	Stirofos
Carbaryl	Malathion	TCMBT
Carbofuran	MCPA	Tebuthiron
Chloroneb	MCPP	Terbacil
Chlorothalonil	Merphos	Terbufos
Chlorpyrifos	Methamidophos	Terbuthylazine
Cyanazine	Methomyl	Terbutryn
2,4-D	Methoxychlor	Toxaphene*
2,4-DB	Metribuzin	Triadimefon
Dazomet	Mevinphos	Trifluralin
DCPA	Nabam	Vapam
DEF	Nabonate	Ziram

* Priority pollutant.

Table 1.7 Technologies applicable to pollutants regulated by pretreatment standards.

Physical	Chemical
Carbon adsorption	Chemical oxidation
Distillation	Chemical precipitation
Filtration	Chromium reduction
Ion exchange	Coagulation
Membranes	Cyanide destruction
Microfiltration (MF)	Dissolved air flotation
Ultrafiltration (UF)	Electrochemical oxidation
Reverse osmosis (RO)	Flocculation
Oil and grease skimming	Hydrolysis
Oil and water separation	Neutralization (pH control)
Sedimentation (clarification)	**Biological**
Steam stripping	Extended aeration
Solvent extraction	Rotating biological contactor
	Sequencing batch reactor
	Trickling filter

*E*FFECTS OF INDUSTRIAL WASTE ON MUNICIPAL WASTEWATER TREATMENT PLANTS

Of some 15 500 municipal wastewater treatment facilities in the U.S., approximately 75% have at least secondary treatment (U.S. EPA, 1987).

Understanding the potential effects of some uncontrolled industrial discharges on WWTP operation will help determine the need for pretreatment in specific situations.

In addition to hydraulic overloads or temperature extremes, potential concerns may include excess amounts of the following:

- Oil and grease;
- Highly acidic or alkaline wastewaters;
- Suspended solids;
- Organic wastes;
- Inorganic wastes;
- Explosive and flammable materials; and
- Wastewater containing volatile, odorous, or corrosive gases.

Because municipal wastewater treatment typically involves complex biochemical processes, an upset affecting one design parameter will likely affect others as well.

Industrial waste characteristics potentially affecting municipal collection systems may include seasonal fluctuations, odorous wastes, corrosiveness, viscous solids precipitation, and explosion hazards. Special care must be exercised to prevent discharge of volatile or flammable chemicals that may result in explosive concentrations in sewer head space.

Industrial-waste-related difficulties likely to affect preliminary or primary treatment facilities may result from overloading or high variability of otherwise acceptable waste materials. Special wastes may result in predictable problems if the WWTP is not designed to handle them. An example may be organic solids from food-processing or meat-packing wastes that become trapped at a point of reduced flow, causing septic odors. This can occur in typical WWTP equipment, such as a grit chamber, or as a result of solids separation equipment plugging.

Unless waste composition is well understood, WWTPs may be underdesigned for not only industrial discharges of solids, including such organics, but also sand or large debris. Well-run industrial facilities, however, can normally collect and dispose of most gross solids, such as rags, as solid waste. A comprehensive cost evaluation may be needed to determine whether the WWTP should address the needed additional capacity or pretreatment is required. It is important for effective operation that actual waste loads and composition be within the design tolerance. As additional dischargers apply to connect to the WWTP, the WWTP staff must constantly reassess potential waste streams and their anticipated effects on treatment effectiveness.

Potential effects on secondary treatment systems may result from materials or circumstances that interfere with, or upset, biochemical activity. These include high variability in waste organic content or flow (shock loading), high organic loading for extended time periods, waste components at potentially toxic levels, nutrient imbalance, and temperature or pH extremes. Tables 1.8 through 1.11 provide U.S. EPA estimates of thresholds of inhibition on biological treatment processes. When waste characteristics are compatible with WWTP processes, shock loadings can be handled with equalization by the responsible waste discharger or by modification of the WWTP, perhaps at the expense of the responsible discharger.

Community user-fee-based solutions often allow for appropriate treatment while optimizing complementary characteristics of different dischargers. An example would be combining alkaline and acidic wastes with a minimum chemical dosage to neutralize pH in the influent. In at least one WWTP agreement, the industrial user assists the WWTP by controlling pH to a level specified daily by the WWTP. This results in reduced WWTP costs and user-fee savings.

Table 1.8 Activated sludge inhibition threshold levels* (U.S. EPA, 1987).

Pollutant	Minimum reported inhibition threshold, mg/L	Reported range of inhibition threshold level, mg/L	Laboratory, pilot or full scale
Metals/nonmetal inorganics			
Cadmium	1	1–10	Unknown
Chromium (total)	1	1–100	Pilot
Chromium (III)	10	10–50	Unknown
Chromium (VI)	1	1	Unknown
Copper	1	1	Pilot
Lead	0.1	0.1–5.0	Unknown
		10–100	Laboratory
Nickel	1	1.0–2.5	Unknown
		5	Pilot
Zinc	0.3	0.3–5	Unknown
		5–10	Pilot
Arsenic	0.1	0.1	Unknown
Mercury	0.1	0.1–1	Unknown
		2.5 as Hg (II)	Laboratory
Silver	0.25	0.25–5	Unknown
Cyanide	0.1	0.1–5	Unknown
		5	Full
Ammonia	480	480	Unknown
Iodine	10	10	Unknown
Sulfide	25	25–30	Unknown
Organics			
Anthracene	500	500	Laboratory
Benzene	100	100–500	Unknown
		125–500	Laboratory
2-Chlorophenol	5	5	Unknown
		20–200	Unknown
1,2-Dichlorobenzene	5	5	Unknown
1,3-Dichlorobenzene	5	5	Unknown
1,4-Dichlorobenzene	5	5	Unknown
2,4-Dichlorophenol	64	64	Unknown
2,4-Dimethylphenol	50	40–200	Unknown
2,4-Dinitrotoluene	5	5	Unknown
1,2-Diphenylhydrazine	5	5	Unknown
Ethylbenzene	200	200	Unknown
Hexachlorobenzene	5	5	Unknown

Table 1.8   Activated sludge inhibition threshold levels* (continued).

Pollutant	Minimum reported inhibition threshold, mg/L	Reported range of inhibition threshold level, mg/L	Laboratory, pilot or full scale
Organics (continued)			
Naphthalene	500	500	Laboratory
		500	Unknown
Nitrobenzene	30	30–500	Unknown
		500	Laboratory
		500	Unknown
Pentachlorophenol	0.95	0.95	Unknown
		50	Unknown
		75–150	Laboratory
Phenathrene	500	500	Laboratory
		500	Unknown
Phenol	50	50–200	Unknown
		200	Unknown
Toluene	200	200	Unknown
2,4,6-Trichlorophenol	50	50–100	Laboratory
Surfactants	100	100–500	Unknown

* References did not distinguish between total of dissolved pollutant inhibition levels.

Table 1.9   Trickling filter inhibition threshold levels* (U.S. EPA 1987).

Pollutant	Minimum reported inhibition threshold, mg/L	Reported range of inhibition threshold level, mg/L	Laboratory, pilot or full scale
Chromium (III)	3.5	3.5–67.6	Full
Cyanide	30	30	Full

* Reference did not distinguish between total or dissolved pollutant inhibition levels.

Table 1.10   Nitrification inhibition threshold levels* (U.S. EPA, 1987).

Pollutant	Minimum reported inibition threshold, mg/L	Reported range of inhibition threshold level, mg/L	Laboratory, pilot or full scale
Metals/nonmetal inorganics			
Cadmium	5.2	5.2	Laboratory
Chromium (T)	0.25	0.25–1.9	Unknown
		1–100	Unkown
		(Trickling filter)	
Chromium (VI)	1	1–10	(as CrO_4^{2-})
Copper	0.05	0.05–0.48	Unknown
Lead	0.5	0.5	Unknown
Nickel	0.25	0.25–0.5	Unknown
		5	Pilot
Zinc	0.08	0.08–0.5	Unknown
Arsenic		1.5	Unknown
Cyanide	0.34	0.34–0.5	Unknown
Chloride		180	Unknown
Organics			
Chloroform	10	10	Unknown
2,4-Dichlorophenol	64	64	Unknown
2,4-Dinitrophenol	150	150	Unknown
Phenol	4	4	Unknown
		4–10	Unknown

* References did not distinguish between total or dissolved pollutant inhibition levels.

When WWTP operations require industrial pretreatment, the user must fully understand the basis and success criteria for pretreatment design. It is little consolation for the user to meet the specified criteria only to find the problem unresolved or the standard not met. Jointly prepared periodic waste load projections are useful in both industrial and WWTP pretreatment planning.

Potential adverse effects on sludge management processes may result from shock loading of normally compatible waste components such as sand, silt, or industrial process waste solids. Perhaps the most important interference of this type is the possibility of sludge contamination, particularly by metals. Sweeping changes in U.S. EPA regulation of sludge disposal may require extensive additional pretreatment. Some industrial wastes may interfere with disinfection processes by, for example, exerting chlorine demand. If accounted for in system design and user fees, this is not likely to be

Table 1.11 Anaerobic digestion threshold inhibition levels[a] (U.S. EPA, 1987).

Pollutant	Minimum reported inhibition threshold, mg/L	Reported range of inhibition threshold level, mg/L	Laboratory, pilot or full scale
Metals/nonmetal inorganics			
Cadmium	20	20	Unknown
Chromium (VI)	110	110	Unknown
Chromium (III)	130	130	Unknown
Copper	40	40	Unknown
Lead	340	340	Unknown
Nickel	10	10	Unknown
		136	Unknown
Zinc	400	400	Unknown
Arsenic	1.6	1.6	Unknown
Silver	13[b]	13–65[b]	Unknown
Cyanide	4	4–100	Unknown
	4	1–4	Unknown
Ammonia	1 500	1 500–8 000	Unknown
Sulfate	500	500–1 000	Unknown
Sulfide	50	50–100	Unknown
Organics			
Acrylonitrile	5	5	Unknown
		5	Unknown
Carbon tetrachloride	2.9	2.9–159.4	Laboratory
		10–20	Unknown
		2.0	Unknown
Chlorobenzene	0.96	0.96–3	Laboratory
		0.96	Unknown
Chloroform	1	1	Unknown
		5–16	Laboratory
		10–16	Unknown
1,2-Dichlorobenzene	0.23	0.23–3.8	Laboratory
		0.23	Unknown
1,4-Dichlorobenzene	1.4	1.4–5.3	Laboratory
		1.4	Unknown
Methylchloride	3.3	3.3–536.4	Pilot
		100	Unknown
Pentachlorophenol	0.2	0.2	Unknown
		0.2–1.8	Laboratory

Table 1.11   Anaerobic digestion threshold inhibition levels[a] (continued).

Pollutant	Minimum reported inhibition threshold, mg/L	Reported range of inhibition threshold level, mg/L	Laboratory, pilot or full scale
Organics (continued)			
Tetrachloroethylene	20	20	Unknown
Tricholoroethylene	1	1–20	Laboratory
		20	Unknown
Trichlorofluoromethane			Unknown

[a] Total pollutant inhibition levels, unless otherwise indicated.
[b] Dissolved metal inhibition levels.

problematic. If, on the other hand, waste components increase over time or fluctuate widely, the WWTP may be unprepared to address needs. As in all need-based cases, consideration of WWTP changes should be weighed against pretreatment options to resolve the concerns.

The most common WWTP concerns relating to industrial waste loading in excess of plant capacity are potential interferences, accumulation, pass-through, and facility damage.

REFERENCE

U.S. Environmental Protection Agency (1987) *1986 Needs Survey Report to Congress: Assessment of Needed Publicly Owned Wastewater Treatment Facilities in the United States.* U.S. EPA 430/9-87-001, Washington, D.C.

SUGGESTED READINGS

Anthony, R.M., and Breimhurst, L.H. (1981) Determining Maximum Influent Concentrations of Priority Pollutants for Treatment Plants. *J. Water Pollut. Control Fed.*, **43**, 10, 1457.

U.S. Environmental Protection Agency (1978) *General Pretreatment Regulations for Existing and New Sources of Pollution.* 43 FR 27746, Washington, D.C.

U.S. Environmental Protection Agency (1986) *Report to Congress on the Discharge of Hazardous Wastes to Publicly Owned Treatment Works.* Office of Water Regulation and Standards, 530-SW-86-004, Washington, D.C.

U.S. Environmental Protection Agency (1991) *National Pretreatment Program: Report to Congress.* EN-336, 21W-4004, Washington, D.C.

Chapter 2
Pretreatment Regulations

The Federal Water Pollution Control Act of 1972, amended in 1977 and 1987, gives the U.S. Environmental Protection Agency (U.S. EPA) the authority to establish and enforce pretreatment standards for the indirect discharge of industrial wastewaters.

The intent of these regulations is to prohibit the discharge of wastes that are incompatible with wastewater treatment plant (WWTP) processes. To be specific, the pretreatment program has three objectives (see 40 CFR Part 403.2):

1. To prevent the introduction of pollutants to WWTPs that will interfere with the operation of a WWTP, including interference with its use or disposal of municipal sludge;
2. To prevent the introduction of pollutants to WWTPs that will pass through the treatment works or otherwise be incompatible with such works; and
3. To improve opportunities to recycle and reclaim municipal and industrial wastewaters and sludge.

Instead of allowing each WWTP to set local limits to regulate contaminants troublesome to its operations, the U.S. Congress passed legislation that requires uniform pretreatment standards to be administered nationwide through authorized pretreatment programs at WWTPs. Pretreatment requirements were focused on controlling a list of 129 (now 126) chemicals that were negotiated in a 1975 lawsuit and became incorporated into Section 307(a) of the Clean Water Act (CWA) as amended in 1977. These pollutants are more commonly known as "priority pollutants."

To implement the mandate of the 1977 CWA to control priority pollutants reaching WWTPs from industrial users, U.S. EPA first issued the General Pretreatment Regulations (40 CFR Part 403) in 1978, followed by a revised version in 1981. These regulations became the basis for nationwide pretreatment programs by establishing procedures, responsibilities, and requirements for U.S. EPA, states, local governments (WWTPs), and industries.

The General Pretreatment Regulations require U.S. EPA to promulgate pretreatment standards for control of priority pollutants in industrial process wastewater before its discharge to the WWTP. U.S. EPA has responded by establishing "prohibited discharge standards" (Part 403.5) applicable to all nondomestic WWTP users and promulgating "categorical pretreatment standards" that are applicable to specific industries (40 CFR Parts 405-471). Congress assigned the primary responsibility for enforcing these standards to local WWTPs.

The Water Quality Act of 1987 reinforced many of the pretreatment provisions of the earlier amendments. While the new law did not change the framework of the general pretreatment regulations, it did establish certain provisions requiring modification to the regulations and mandated that U.S. EPA report to Congress on pretreatment (which U.S. EPA sent to Congress on May 21, 1991). Perhaps the most significant provision from an industrial user standpoint dealt with "removal credits."

The Resource Conservation and Recovery Act (RCRA) Section 3018(b) contains a provision dealing with Domestic Sewage Exclusion and a required study. On February 7, 1986, U.S. EPA submitted its "Report to Congress on the Discharge of Hazardous Waste to Publicly Owned Treatment Works." As a follow-up commitment made within that report, U.S. EPA promulgated changes to the General Pretreatment Regulations (see 55 FR 30082) on July 24, 1990. These changes have a significant effect on industrial users.

Readers should obtain copies of the relevant changes to the statute for an in-depth understanding.

Industrial users (IU) of WWTPs with authorized pretreatment programs must meet certain requirements before industrial wastewater is allowed to be discharged to a WWTP for treatment. The regulatory framework for these requirements is set forth in 40 CFR Part 403, "General Pretreatment Regulations for Existing and New Sources of Pollution."

Generally, pretreatment requirements apply to nondomestic dischargers (that is, IUs) who are covered by pretreatment standards and whose flows enter WWTPs either through the WWTP's sewer system or by truck or rail, or are otherwise introduced to the WWTPs. The current regulations define "significant industrial user" (40 CFR 403.3[t]) as any IU covered by a categorical pretreatment standard. Also, an IU may be considered significant if its process wastewater flow exceeds 95 m^3/d (25 000 gpd) or meets other conditions. Even if an IU is not considered a significant industrial user, certain requirements still apply (for example, hazardous waste notification at 40 CFR 403.12[p]). To determine if an IU is covered by a categorical pretreatment standard, the IU should refer to 40 CFR 403.6 and the appropriate subchapters of 40 CFR Chapter 1 (see Table 2.1).

SIGNIFICANT INDUSTRIAL USER REQUIREMENTS

Once an IU is classified as a significant user, the following may apply.

The WWTP (see 40 CFR 403.8[f][6]) is required to prepare a list of significant IUs and, within 30 days of approval of the list by the approval authority (typically the regional U.S. EPA office), the WWTP is to notify each significant IU of its status and the requirements it must implement (see 40 CFR 403.8[f][2][iii]).

The requirement (see 40 CFR 403.8[f][2][v]) states that an IU will be inspected on an annual basis by the WWTP and have its effluent sampled. Several other requirements, such as an evaluation of the IU's control plan for slug discharges, are also within this requirement.

The regulations (see 40 CFR 403.8[f][1][iii]) state that significant IUs must have a permit or equivalent individual control mechanism with their WWTPs, with durations of no longer than 5 years. An equivalent control mechanism is one that has the same degree of specificity and control as a permit. U.S. EPA pointed out in the preamble to the July 24, 1991, final rule that an ordinance or contract does not constitute an equivalent individual control mechanism. In addition, the preamble also states that IUs are not required to obtain a permit before discharging to a WWTP, but that they must comply with all applicable pretreatment requirements under federal law, including those not contained in a permit or equivalent individual control mechanism. Finally, compliance by an IU with the terms of the permit does *not* shield it from liability for failure to comply with federal pretreatment requirements not set forth in the permit.

Table 2.1   Industries with categorical pretreatment standards.[a]

Industry category	40 CFR part	Relevant SIC code(s)[b]	Pretreatment standards	
Aluminum forming	467	3353,3354,3355,3357,3363	ES[c]	NS[d]
Asbestos manufacturing	427	2621,3292	—	NS[e]
Battery manufacturing	461	3691,3692	ES	NS
Builders' paper and board mills	431	2621	ES	NS
Carbon black manufacturing	458	2895	—	NS
Cement manufacturing	411	3241	—	NS[f]
Coil coating canmaking	465	3479,3492,3411	ES	NS
Copper forming	468	3351,3357,3463	ES	NS
Dairy products processing	405	2021,2022,2023,2024,2026	—	NS[g]
Electrical and electronic components	469	3671,3674,3679	ES	NS
Electroplating	413	3471,3672	ES	NS
Feedlots	412	0211,0213,0214	—	NS[h]
Ferroalloy manufacturing	424	3313	—	NS[i]
Fertilizer manufacturing	418	2873,2874,2875	—	NS
Fruits and vegetables processing	407	2033,2034,2035,2037	—	NS[j]
Glass manufacturing	426	3211,3221,3296	—	NS[k]
Grain mills manufacturing	406	2041,2043,2044,2045 2046,2047	—	NS[l]
Ink formulating	447	2893	—	NS
Inorganic chemicals	415	2812,2813,2816,2819	ES	NS
Iron and steel manufacturing	420	3312,3315,3316,3317,3479	ES	NS
Leather tanning and finishing	425	3111	ES	NS
Metal finishing	433	Industry groups: 34,35,36,37,38	ES	NS
Metal molding and casting (foundries)	464	3321,3322,3324,3325 3365,3366,3369	ES	NS
Nonferrous metals forming and metal powders	471	3356,3357,3363,3497	ES	NS
Nonferrous metals manufacturing	421	2819,3331,3334,3339,3341	ES	NS
Organic chemicals, plastics, and synthetic fibers	414	2821,2823,2824,2865,2869	ES	NS[m]
Paint formulating	446	2851	—	NS[n]
Pesticide chemicals manufacturing formulation and packaging	455	2879	ES[o] —[p]	NS[o] —[p]
Petroleum refining	419	2911	ES	NS
Pharmaceuticals manufacturing	439	2833,2834	ES[q]	NS[q]
Porcelain enameling	466	3431,3631,3632,3633,3639 3469,3479	ES	NS
Pulp, paper, and paperboard	430	2611,2621,2631	ES[r]	NS[r]
Rubber manufacturing	428	2822	—	NS[s]

Table 2.1 Industries with categorical pretreatment standards[a] (continued).

Industry category	40 CFR part	Relevant SIC code(s)[b]	Pretreatment standards	
Soaps and detergents manufacturing	417	2841	—	NS[t]
Steam electric power generation	423	4911	ES	NS
Sugar processing	409	2061,2062,2063	—	NS[u]
Timber products processing	429	2491,2493	ES	NS

[a] Of the categories subject to pretreatment standards (Part 403, Appendix C), only categories with *numerical* pretreatment standards are shown. The other categories simply contain references to Part 128 (now deleted from 40 CFR), Part 403 (the General Pretreatment Standards), or to removal credits (less stringent standard based on WWTP commitment to remove a specified percentage of "incompatible" pollutants). These stock passages offer no basis for setting numerical limits for specific pollutants in industrial user control mechanisms.

[b] Standard Industrial Classification. These codes were taken from the 1987 SIC Manual, which may be ordered as Publication No. PB 87-100012 from the National Technical Information Service (NTIS), 5285 Port Royal Road, Springfield, VA 22161.

[c] Existing Sources (PSES).

[d] New Sources (PSNS). New sources are regulated processes for which construction *started after* the date PSNS was proposed. See Part 122.2 for the distinction between a "new source" and a "new discharger."

[e] Subparts A(427.15), B(427.25), C(427.35), D(427.45), E(427.55), F(427.65), and G(427.75).

[f] Subparts A(411.15), B(411.25), and C(411.35)

[g] Subparts A(405.15), B(405.25), C(405.35), D(405.45), E(405.55), F(405.65), G(405.75), H(405.85), I(405.95), J(405.105), K(405.115), and L(405.125).

[h] Subparts A(412.15) and B(412.15).

[i] Subparts A(424.15), B(424.25), and C(424.35).

[j] Subparts A(407.15), B(407.25), C(407.35), D(407.45), and E(407.55).

[k] Subparts A(426.15), B(426.25), C(426.35), D(426.45), E(426.55), G(426.75), K(426.116), L(426.126), and M(426.136).

[l] Subparts A(406.16), B(406.25), C(406.35), D(406.45), E(406.55), and F(406.65).

[m] Revised PS scheduled for 1993. FR 57 at 19751 (May 7, 1992).

[n] Subpart A(446.16).

[o] Proposed in 1992. Final PS schedule for 1993. FR 57 at 19751.

[p] Scheduled to be proposed in 1994. FR 57 at 19751.

[q] Revised PS scheduled for proposal in 1994. FR 57 at 19751.

[r] Revised PS scheduled for proposal in 1993. FR 57 at 19751.

[s] Subparts E(428.56), F(428.66), G(428.76), J(428.106), and K(428.116).

[t] Subparts O(417.156), P(417.166), Q(417.176), and R(417.186).

[u] Subpart A(409.15).

PROHIBITIONS

The General Pretreatment Regulations (see 40 CFR 403.5) list several general and specific prohibitions. In general, an IU pollutant discharge to a WWTP may not cause interference with or pass through the WWTP or interfere with the WWTP's sludge disposal options.

In addition, eight specific prohibitions require IU compliance. According to 40 CFR 403.5(b), the following eight pollutants may not be introduced to a WWTP:

(1) Pollutants that create a fire or explosion hazard in the municipal WWTP, including, but not limited to, waste streams with a closed cup flashpoint of less than or 60°C (140°F) using the test methods specified in 40 CFR 261.21.

(2) Pollutants that will cause corrosive structural damage to the municipal WWTP (but in no case discharges with pH lower than 5.0) unless the WWTP is specifically designed to accommodate such discharges.

(3) Solid or viscous pollutants in amounts that will cause obstruction to the flow in the WWTP resulting in interference.

(4) Any pollutant, including oxygen-demanding pollutants (such as BOD), released in a discharge at a flow rate and/or pollutant concentration that will cause interference with the WWTP.

(5) Heat in amounts that will inhibit biological activity in the WWTP resulting in interference, but in no case heat in such quantities that the temperature at the WWTP exceeds 40°C (104°F) unless the approval authority, upon request of the POTW, approves alternate temperature limits.

(6) Petroleum oil, nonbiodegradable cutting oil, or products of mineral oil origin in amounts that will cause interference or pass through.

(7) Pollutants that result in the presence of toxic gases, vapors, or fumes within the POTW in a quantity that may cause acute worker health and safety problems.

(8) Any trucked or hauled pollutants, except at discharge points designated by the POTW.

Where a WWTP determines that either the general or specific prohibitions do not protect it, the WWTP can set and enforce local limits on pollutants or pollutant parameters as it deems necessary. Such local limits are considered pretreatment standards for purposes of enforcement.

CATEGORICAL INDUSTRIAL USERS

When U.S. EPA promulgates categorical pretreatment standards that affect an IU, the IU has certain reporting, recordkeeping, and other obligations. These requirements are set forth in 40 CFR 403.12, as they relate to certain reporting requirements that must be met on specific timetables. The general requirements are listed below.

Industrial users must file a baseline monitoring report (BMR) within 180 days after the categorical pretreatment standard's effective date. The IU furnishes an initial process description and statement certifying that it is in compliance with the categorical pretreatment standards or, if not, that it will adhere to a schedule bringing it in compliance by the applicable date (40 CFR 403.12[b]).

The IU must submit progress reports within 14 days of each milestone date on the compliance schedule submitted with the BMR. Each such report must be filed within 9 months of the prior progress report (40 CFR 403.12[c]) and each compliance schedule milestone must be no longer than 9 months apart.

At the time the IU comes into compliance with the categorical pretreatment standards, such date generally being 3 years after the promulgation of the applicable categorical standard, it is then required within 90 days of the compliance date to file a "90 day compliance report." This report must certify that the compliance has been achieved and contain appropriate monitoring data showing compliance (40 CFR 403.12[d]).

In June and December of each year, the IU files reports on continued compliance containing various monitoring results and other information (40 CFR 403.12[e]).

In addition to the above, 40 CFR 403.12 contains other significant reporting requirements. These include slug loading notifications, signatory requirements, recordkeeping requirements, and others.

SIGNIFICANT NONCATEGORICAL INDUSTRIAL USERS

The regulations require that significant noncategorical IUs file special semi-annual reports even if the IU is not covered by a categorical pretreatment standard (see 40 CFR 1403.12[h]). In other words, once an IU is a significant user, the 6-month reports must be filed, containing much of the same information specified above for categorical IUs.

Hazardous Waste Notifications

On July 24, 1990, the revisions to the General Pretreatment Regulations added a new section, 40 CFR 403.12(p), requiring that all IUs notify each WWTP, the U.S. EPA Regional Waste Management Division Director, and the state hazardous waste authority of any discharge to the WWTP of a substance that, if otherwise disposed of, would be a hazardous waste under 40 CFR Part 261. The initial notification of hazardous waste discharges was to be made by February 19, 1991, or within 180 days of commencement of the discharge. The requirements contain specific direction as to what the notification is to include as well as a certification statement that must be signed.

Removal Credits

The General Pretreatment Regulations contain provisions dealing with removal credits. Simply stated, if a WWTP can treat certain pollutants to high levels of removal, it can obtain permission from U.S. EPA (or a state that has been granted authority by U.S. EPA) to issue credit to IUs, providing the IU with less stringent pretreatment requirements for that pollutant. Removal credits for several pollutants are available through 40 CFR Part 403 (Appendix G-1) when disposal practices used by a WWTP meet the requirements of 40 CFR Part 503 for the intended practice (land application, surface disposal, or incineration). Additionally, when a pollutant listed in 40 CFR Part 403, Appendix G-II, does not exceed the concentration listed in the table, a removal credit can be obtained (see Tables 2.2 and 2.3). Where a WWTP sends all of its wastewater solids to a municipal solid waste landfill covered by 40 CFR Part 258 and the WWTP can demonstrate that the landfill is in compliance with Part 258, it can then apply for removal credit authority.

Other Provisions

The General Pretreatment Regulations also contain several other sections (see 40 CFR 403.8 and 403.9) that should be studied carefully by IUs. These include provisions placed on WWTPs to establish and obtain approval for "pretreatment programs." While this is predominantly a requirement that the WWTP must exercise, it does have certain spinoff requirements placed on the IU once approved. In addition, U.S. EPA has prepared a series of guidance materials on the preparation of pretreatment programs (as well as other issues).

In applying the categorical pretreatment standards and determining which limitations must be achieved by the IU at the point that its flow enters the

Table 2.2   Regulated pollutants in Part 503 eligible for a removal credit (40 CFR Part 403.7).

Pollutants	Use or disposal practice		
	LA[a]	SD[b]	I[c]
Arsenic	X	X	X
Beryllium			X
Cadmium	X		X
Chromium	X	X	X
Copper	X		
Lead	X		X
Mercury	X		X
Molybdenum	X		
Nickel	X	X	X
Selenium	X		
Zinc	X		
Total hydrocarbons			X[d]

[a] LA = land application.
[b] SD = surface disposal site without a liner and leachate collection system.
[c] I = firing of wastewater sludge in a wastewater sludge incinerator.
[d] The following organic pollutants are eligible for a removal credit if the requirements for total hydrocarbons in subpart E in 40 CFR Part 503 are met when sludge is fired in an incinerator: acrylonitrile, aldrin/dieldrin (total), benzene, benzidine, benzo(a)pyrene, bis(2-chloroethyl)ether, bis(2-ethylhexyl)phthalate, bromodichloromethane, bromoethane, bromoform, carbon tetrachloride, chlordane, chloroform, chloromethane, DDD, DDE, DDT, dibromochloromethane, dibutyl phthalate, 1,2-dichloroethane, 1,1-dichloroethylene, 2,4-dichlorophenol, 1,3-dichloropropene, diethyl phthalate, 2,4-dinitrophenol, 1,2-diphenylhydrazine, di-n-butyl phthalate, endosulfan, endrin, ethylbenzene, heptachlor, heptachlor epoxide, hexachlorobutadiene, α-hexachlorocyclohexane, β-hexachlorocyclohexane, hexachlorocyclopentadiene, hexachloroethane, hydrogen cyanide, isophorone, lindane, methylene chloride, nitrobenzene, N-nitrosodimethylamine, N-nitrosodi-n-propylamine, pentachlorophenol, phenol, polychlorinated biphenyls, 2,3,7,8-tetrachlorodibenzo-p-dioxin, 1,2,2,2,-tetra- chloroethane, tetrachloroethylene, toluene, toxaphene, trichloroethylene, 1,2,4-trichlorobenzene, 1,1,1-trichloroethane, 1,1,2-trichloroethane, and 2,4,6-trichlorophenol.

Table 2.3   Additional pollutants eligible for a removal credit (mg/kg, dry weight basis).

Pollutant	LA[a]	Unlined[b]	Lined[c]	I[d]
		Use or disposal practice		
		Surface disposal		
Arsenic	—	—	100	—
Aldrin/dieldrin (total)	2.7	—	—	—
Benzene	16[e]	140	3 400	—
Benzo(*a*)pyrene	15	100[e]	100[e]	—
Bis(2-ethylhexyl)phthalate	—	100[e]	100[e]	—
Cadmium	—	100[e]	100[e]	—
Chlordane	86	100[e]	100[e]	—
Chromium	—	—	100[e]	—
Copper	—	46[e]	100[e]	1 400
DDD, DDE, DDT (total)	1.2	2 000	2 000	—
2,4-Dichlorophenoxy-acetic aid	—	7	7	—
Fluoride	730	—	—	—
Heptachlor	7.4	—	—	—
Hexachlorobenzene	29	—	—	—
Hexachlorobutadiene	600	—	—	—
Iron	78[e]	—	—	—
Lead	—	100[e]	100[e]	—
Lindane	84	28[e]	28[e]	—
Malathion	—	0.63	0.63	—
Mercury	—	100[e]	100[e]	—
Molybdenum	—	40	40	—
Nickel	—	—	100[e]	—
N-Nitrosodimethylamine	2.1	0.088	0.088	—
Pentachlorophenol	30	—	—	—
Phenol	—	82	82	—
Polychlorinated biphenyls	4.6	<50	<50	—
Selenium	—	4.8	4.8	4.8
Toxaphene	10	26[e]	26[e]	—
Trichloroethylene	10[e]	9 500	10[e]	—
Zinc	—	4 500	4 500	4 500

[a] LA = land application.
[b] Wastewater sludge unit without a liner and leachate collection system.
[c] Wastewater unit with a liner and leachate collection system.
[d] I = incineration.
[e] Value expressed in grams per kilogram—dry weight basis.

WWTP, a provision entitled the "Combined Waste Stream Formula" (see 40 CFR 403.6[e]) should be carefully reviewed. In addition, other sections of 40 CFR 403.6 dealing with deadlines, concentration and mass limits, and choice of monitoring locations are also important.

Rules are set forth in 40 CFR 403.13 dealing with variances from categorical pretreatment standards for fundamentally different factors. While to date this has had little or no use, it is available in the event an IU initiates a timely filing and makes the appropriate demonstrations.

Four other sections worthy of consideration include 40 CFR 403.14 (Confidentiality), 40 CFR 403.15 (Net/Gross Calculations), 40 CFR 403.16 (Upset Provision), and 40 CFR 403.17 (Bypass). Careful review of these provisions is important to ensure that enforceable requirements are met and that proprietary information can adequately be protected.

Because the purpose of this manual is to provide general design guidance and facility development information for the pretreatment of industrial wastewater, the underlying regulations should be carefully reviewed by each IU to ensure the requirements are being met. Most WWTPs with significant IU loads will be required to have local pretreatment programs; therefore, it is further recommended that IUs obtain copies of these pretreatment programs from their local WWTPs, study them carefully, and make certain they are complying with any requirements applicable to them. While certain general regulations cover the ways local pretreatment programs must be developed by WWTPs, significant differences exist between such programs to the extent that an overview discussion at this point is not practical.

FUTURE OUTLOOK

The above discussions center on the national pretreatment program existing at present. It is difficult to predict what new requirements will be added to the program in ensuing years. Such changes may be driven by additional legislative action to amend the Clean Water Act, as well as several current U.S. EPA programs to complete requirements mandated by prior amendments. It is recommended that the users of this manual supplement the references quoted herein with any new U.S. EPA regulations in existence at the time design and facility development activities are undertaken.

Chapter 3
Management Strategies for Pollution Prevention and Waste Minimization

This chapter provides an overview of the alternatives available for managing the discharge of industrial wastewater. The goal of any pretreatment program is to manage the waste streams in a manner that ensures regulatory requirements are consistently and cost-effectively met. This chapter describes general, technical, and economic considerations and principles that should be addressed and evaluated before implementing any wastewater discharge management strategy. Figure 3.1 shows the key components of a strategy for wastewater management at a manufacturing facility. The critical factor for successful development and implementation of a management strategy is the continuous commitment of everyone involved within the company, beginning with top management, to support the objectives of the agreed strategy. Management must recognize the benefits of pollution prevention and waste minimization and integrate these practices into the corporate or plant philosophy.

*I*N-PLANT SURVEY

In developing a wastewater management strategy, it is first necessary to have detailed information on the wastewater generated within the facility. In-plant surveys generate data on the sources, chemical compositions, quantities, variations, distribution, and discharge frequencies and durations of all industrial process waste streams. These data are used to establish a baseline description of the wastewater produced at the facility and develop or model possible management strategies. Additionally, the baseline data provide a starting point for considering the effect of future production growth, water conservation efforts, or changing regulatory requirements.

The most important aspect of conducting an in-plant survey is making it as complete as possible. This requires a full understanding of all production activities within the facility and detailed, accurate drawings of the plant showing the locations of the various processing units, their water distribution, and wastewater production and collection systems. The waste stream from each unit process is studied, measured, and analyzed to determine the frequency, quantity, and flow rate of the discharge and the nature and concentration of the pollutants present. If possible, flow measurements and sample collections should be performed using permanent monitoring stations. If these are not available, temporary data collection points should be used. The frequency, extent, and type of monitoring and sampling needed depend on the nature and variability of each waste stream. To ensure that the characterization of each waste stream is complete and representative, it is often appropriate to prepare an advance sampling and monitoring plan to guide the study.

Once all individual waste streams have been fully characterized, they can be sorted and grouped according to the types and/or concentrations of pollutants present, or the applicability of U.S. Environmental Protection Agency (U.S. EPA) categorical pretreatment standards. During the survey, raw material use and production records should be obtained. Correlations between production, or material use, and waste generation can then be developed for future decision making.

IDENTIFYING CATEGORICAL WASTE STREAMS. Any waste streams covered by categorical pretreatment standards should further be

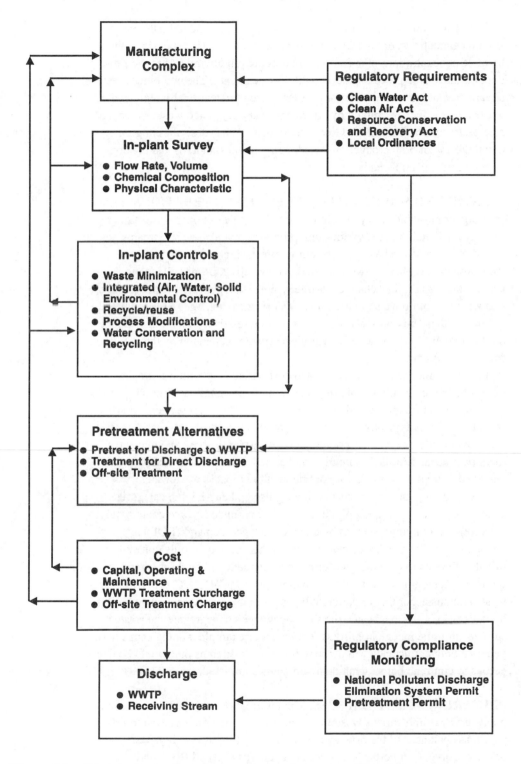

Figure 3.1   Strategy for wastewater management at a manufacturing complex.

identified as subject to either production-based standards, combined waste stream calculations, or both. In the case of production-based standards, the allowable concentrations of specific pollutants are directly related to the production rate of the process generating the waste stream. The combined waste stream formula is used to determine compliance of categorically regulated waste streams that are diluted by noncategorically regulated waste streams before pretreatment. U.S. EPA's Guidance Manual (1985) describes in greater detail the use of both production-based pretreatment standards and the combined waste stream formula.

IDENTIFYING WASTEWATER-GENERATING OPERATIONS. In performing the in-plant survey, it is important to include not only waste streams directly attributable to the various unit processes, but also waste streams generated from cross-media pollution control efforts, including waste streams from wet air scrubbers blowdown, sludge dewatering, or miscellaneous sources such as product change washouts, site cleanup efforts, yard drainage, noncontact cooling water, or secondary containment spillage. Although the volume of these auxiliary waste streams may be relatively small, their effect on the treatability of the total industrial wastewater generated by the facility may be significant.

In light of increasing restrictions and costs associated with off-site disposal of spent concentrated plating or solvent baths, many facilities are considering on-site treatment of these wastes. Although discharges generated from this activity typically are infrequent or on a slug-discharge basis, they must be considered and incorporated into the in-plant survey and management plan because of their potential to overload any treatment system that is designed to treat only regular or continuous flow process wastewater.

In conducting an in-plant survey, it may also be beneficial to categorize waste streams according to pollutant types to determine if some contain only conventional pollutants such as biochemical oxygen demand (BOD) and suspended solids. In many instances, waste streams containing only conventional pollutants may require no further pretreatment and be discharged without treatment to a wastewater treatment plant (WWTP). Classifying waste streams according to pollutant types also may expose any incompatibility between waste streams and the subsequent need to segregate the waste streams until the incompatibility is removed. For example, plating shops may generate both acidic and cyanide-containing waste streams that would be dangerous to combine before cyanide treatment is completed.

PREPARING MASS BALANCES. The information obtained from the in-plant survey of the various wastewater-generating operations is used to prepare mass balances of the flow and waste loads within the facility. Mass balances serve as important cross checks to confirm that all flows and pollutant loads have been accounted for in the survey and that no major waste stream or pollutant-generating operations have been missed.

Properly prepared mass balances provide data that can be used in preparing a treatment strategy and determining flow equalization requirements. They also serve as indicators of potential areas where waste minimization or pollution prevention practices should be considered or implemented.

Developing flow and waste load balance for an entire facility can be a complex undertaking, especially if there are multiple wastewater-generating processes that operate independently of each other and flow through a combined wastewater collection system. However, it is often desirable to prepare individual mass balances for each wastewater-generating process. Each individual mass balance is prepared by accounting, as accurately as possible, for the conversion of raw materials into products, solid wastes, wastewater loadings, and so on over a specific period of time. Because water often serves as both a raw material and a cleanup agent, its consumption and fate must be monitored closely during the mass balance period, and wastewater flows, product incorporation, and evaporative losses must be measured accurately. The individual mass balances, once prepared, can be consolidated into an overall mass balance for the entire facility. If possible, the results of the combined mass balance should be verified by measuring the flow of the total effluent from the facility over a representative period of time while simultaneously analyzing it for pollutants of concern.

*I*N-PLANT CONTROL

Once the mass balance for the facility is completed and the sources and loadings of the various waste streams have been determined, it is appropriate to consider options for the control and reduction of pollutants to reduce the concentrations and volumes of the waste streams that need pretreatment.

Ideally, the first choice is to eliminate the generation of specific pollutants by substituting different raw materials that generate no wastewater or only wastewater that is compatible with the WWTP and requires no pretreatment. Because it is often impossible or economically infeasible to eliminate pollutant-generating raw materials from the production process as a result of product specifications or other reasons, the possibility of recycling or reusing the wastewater generated during the production process should be evaluated and explored. In some instances, the concentrated solutions obtained from the cleanup operations can be recycled as part of the starting materials for the next production run. If it is not feasible to recycle generated wastewater internally, the possibility of having an outside party either reclaim or reuse the wastewater should be explored.

If efforts to either eliminate, recycle, or reclaim wastewater through changes in production activities are unsuccessful, steps should be taken to reduce the quantity of wastewater that requires treatment. These steps include implementing good housekeeping practices, using spill-control measures such as spill-containment enclosures and drip trays around tanks; eliminating any "wet floor" areas; and using rinses that are either static or do not have any overspray. Proper housekeeping should be practiced at all times, because it can be one of the most cost-effective measures for reducing pollutant loadings and maintaining compliance with regulatory requirements.

A facility's pollution prevention and waste minimization effort should be continuous rather than an isolated activity. For this type of program to succeed, specific measurable goals should be established and communicated to everyone at the facility. All successes should be recognized and publicized. Without continuous commitment and support from all levels of staff to achieve the goals of waste minimization through raw materials substitution, process modification, recycling (wastewater segregation and reuse), reclamation, and good housekeeping practices, chances for the long-term success of the program are decreased and any "significant" achievements may be only temporary. The management strategy for the control and treatment of a facility's wastes needs to be incorporated at the beginning of the plan and linked with all other components of the planning and implementation process. Benefits of a well-implemented plan include lower costs, improved product quality, increased production, improved public relations, reduced liability, and successful regulatory compliance.

WATER CONSERVATION AND RECYCLING. Efforts to conserve and recycle water should be incorporated into a waste minimization program or initiated as a separate activity with its own specific goals. The benefits of reducing the volume of wastewater discharges through recycling, reuse, and other conservation methods may include lower capital costs for the construction of a smaller pretreatment system and lower operating costs for the system because of possible reductions in the variability of the waste stream being treated. Additionally, smaller user charges for discharging to the WWTP can result if the WWTP's fees include a volume component. However, simple reduction of water alone, in the absence of other measures, may do little to reduce the ultimate treatment costs unless increased concentration results in increased treatment efficiency. Actually, the higher pollutant concentrations obtained when reducing only the volume of a waste stream may increase the risk that the treatment capability of the system is exceeded, thereby resulting in discharge violations. Because there are instances where it is more efficient to treat concentrated streams rather than dilute them, it is important to investigate the effect that any concentration change resulting from water conservation may have on the treatment process.

Water conservation alternatives include, but are not limited to, the reuse of cooling water as product makeup or cleanup water, collecting stormwater for noncritical water uses, flow-restricting or water-saving devices, and the recycling of water in closed-loop systems. Once all internal applications for waste reuse, recycling, and conservation are maximized and implemented, the possibility remains that treated wastewater, if suitable, can be used by outside contractors for irrigation, dust control, or other tasks that normally would use fresh water. If the treated effluent is discharged to a WWTP, consideration should be given to the presence of any potential pollutants that may interfere with the WWTP's ability to reclaim its treated wastewater or use it for wetlands reclamation projects.

PRETREATMENT

Pretreatment of industrial wastewater before discharge to a WWTP may be necessary for several reasons. As described in Chapter 2, some industries are subject to federal or local pretreatment standards because they discharge organic and/or inorganic pollutants that can damage conveyance systems, cause inhibitory effects or pass through the WWTP treatment processes, or interfere with selected sludge-disposal alternatives. Other industries may voluntarily choose pretreatment to reduce or avoid WWTP surcharges on pollutants such as BOD and suspended solids. In rare instances, the wastewater solids or residuals resulting from the pretreatment process may be valuable in their own right, such as when they contain significant amounts of precious metals. Pretreatment systems can be installed to assist with the reclamation of these materials.

The type of pretreatment selected, be it physical, chemical, or biological, depends on wastewater characteristics, applicable pretreatment standards, and anticipated production changes that may affect wastewater characteristics. Physical treatment methods are designed primarily to remove suspended solids, settleable solids, and oil and grease. Chemical treatment typically removes dissolved and colloidal solids, nutrients, heavy metals, and similar pollutants. Finally, biological treatment is typically used to remove biodegradable organics and nutrients in industrial wastewater. The processes most often used for industrial wastewater treatment are provided in Table 3.1.

Before selecting pretreatment options, industries must consider several factors that will affect the decision-making process. U.S. EPA has issued numerous pretreatment standards relative to specific industrial categorical wastewater discharges for both existing sources and new sources. In addition, states and municipalities have frequently supplemented the federal standards with pretreatment requirements that take into account local conditions. Conditions such as stringent water quality standards reflected in the WWTP's National Pollutant Discharge Elimination System (NPDES) permit or the WWTP's treatment process characteristics may require more restrictive industrial discharge limits in some cases. The selection of pretreatment options must guarantee that the constructed pretreatment facility, as well as its

Table 3.1   Processes applicable to industrial wastewater treatment.

Pollutant	Processes
Biological oxygen demand (biodegradable organics)	Aerobic biological: activated sludge, aerated lagoons, trickling filters, rotating biological contactors, oxidation ditches, stabilization basins, packed bed reactors Anaerobic biological: anaerobic lagoons, anaerobic filters, anaerobic contact, fluidized bed reactors
Total suspended solids	Sedimentation, flotation, screening, filtration, coagulation/flocculation/ sedimentation or flotation
Refractory organics (COD, TOC)	Carbon adsorption, chemical oxidation
Nitrogen	Ammonia stripping, nitrification and denitrification, ion exchange, breakpoint chlorination
Phosphorus	Precipitation, biological uptake, ion exchange
Heavy metals	Membrane filtration, evaporation and electrodialysis, chemical precipitation, ion exchange
Dissolved inorganic solids	Ion exchange, reverse osmosis, electrodialysis
Oil and grease	Coagulation/flocculation/flotation, ultrafiltration
Volatile organic compounds	Aeration, chemical oxidation adsorption, biological treatment

discharge, will comply with all regulatory requirements. Additionally, long-term considerations should be addressed in the selection process, including provisions for additional treatment in the future to meet changing regulatory requirements or for the addition of modular systems to account for long-term flow variations. In some instances, field-scale pilot testing, trial uses of modified production processes, or research and development must be conducted before implementation of a pretreatment program.

PHYSICAL SEPARATION. Physical separation processes typically include flow equalization, screening, sedimentation, flotation, filtration, aeration, and adsorption. A brief description of each physical treatment method and a discussion of the advantages and disadvantages follow. More detail is found in Chapter 5.

Screening is used to remove coarse solids (rags or pieces of wood, for example) and prevent damage to or clogging of downstream equipment. Manually cleaned screens work well, but cleaning them requires labor and may cause overflows from clogging. Mechanically cleaned screens also perform well, but they may become jammed because of obstructions such as bricks or pieces of wood.

Flow equalization dampens flow variation and thus achieves a fairly constant flow rate to the sewer system. It also dampens the concentration and mass flow of wastewater constituents, yielding a more uniform loading to the WWTP. Flow equalization helps reduce shock hydraulic, organic, and nutrient loads and can reduce the required size of pretreatment facilities.

Sedimentation is the removal of suspended solids by gravity separation in a quiescent basin. Typically, sedimentation is highly reliable. However, the sludge collector mechanism may occasionally jam. The proper design of bottom slope and scraper blades and the appropriate number of arms will reduce this problem. Surface scum may cause odors that can be controlled by frequent removal. Short circuiting and poor performance may occur if inlet and outlet designs are inadequate.

Dissolved air flotation (DAF) is used to remove suspended solids by causing them to rise to the surface. One DAF process consists of saturating some or all of the wastewater feed or a portion of recycled effluent with air under pressure. The pressurized wastewater is held for up to 3 minutes in a retention tank and then released to atmospheric pressure in the flotation chamber. Upon exposure to atmospheric pressure, microscopic air bubbles are released which attach to oil and suspended particles, floating these particles to the surface where they are skimmed off as float solids. Dissolved air flotation systems are reliable, but chemical addition is often used to enhance performance. These systems require little land area, but air compressor noise must be controlled, and the sludge must be treated and receive proper disposal (Viessman and Hammer, 1985).

Filtration is a solid–liquid separation process in which the liquid passes through a porous medium to remove fine suspended solids. Filtration is reliable and requires little use of land. However, backwash water will need to be treated, resulting in the production of solids, which require disposal.

Aeration may be used to strip volatile compounds from industrial wastewater. Diffused aeration or mechanical aeration typically are used. The aeration process is simple in concept and typically reliable. Generally, land requirements are small and sludge is not generated in an aeration system designed simply for aeration and not biological treatment. Proper design must ensure that offgases do not cause air pollution problems.

Adsorption is the accumulation of a substance at the surface of a solid material (usually activated carbon) called an adsorbent. Carbon systems generally consist of vessels in which granular carbon is placed, forming a filter bed through which wastewater passes. Adsorption systems require little land.

Under anaerobic conditions, biological activity in carbon beds may generate hydrogen sulfide, which has an unpleasant odor. Spent carbon may create a land disposal problem, unless regenerated. However, regeneration systems are expensive and may cause air pollution. Granular carbon systems often require pretreatment to reduce solids loadings to the beds. Powdered carbon may be used instead of granular carbon, but typically it is fed to wastewater using chemical feed equipment rather than being contained in a bed or column (Weber, 1972).

The three major membrane processes are reverse osmosis, electrodialysis, and ultrafiltration. Reverse osmosis is the pressurized transport of a solvent across a semipermeable membrane that impedes passage of solute (pollutants) but allows solvent (usually water) flow. Fouling of the membrane may result from the deposition of colloidal or suspended materials in the wastewater; thus, pretreatment typically is required to avoid frequent cleaning requirements. Chemical recovery and wastewater reuse are possible.

In electrodialysis (a physical-chemical process), an electric current induces partial separation of wastewater components. The separation is achieved by alternately placing cation and anion selective membranes across the current path. When current is applied, the cations pass through the cation exchanger membrane in one direction, and the anions pass through the anion exchanger membrane in the other direction. Chemical recovery and wastewater reuse are possible, but power costs are typically high and membrane fouling may be a problem.

For ultrafiltration, wastewater is pumped past a membrane. Under the applied pressure, water and most dissolved constituents pass through the pores of the membrane, while larger molecules such as colloids and emulsified oils are retained. The process typically has high capital and operations and maintenance costs, and membrane fouling may be a problem. However, it seems to be a reliable technology for certain applications.

CHEMICAL PRETREATMENT. Chemical pretreatment processes typically include pH neutralization, chemical precipitation, oxidation-reduction, and ion exchange. Each type of chemical treatment and its advantages and disadvantages are discussed below.

Neutralization involves the addition of acids or bases to wastewater for adjusting pH to an allowable range, typically pH 5 to 9. Acidic wastewaters typically are neutralized with lime ($Ca(OH)_2$), caustic soda ($NaOH$), or soda ash (Na_2CO_3). Slaked lime is often used because it is less expensive than $NaOH$ and Na_2CO_3. Sodium hydroxide is also sometimes preferred because of its lower maintenance requirements and ease of use. Alkaline wastewaters are typically neutralized with sulfuric acid, hydrochloric acid, or carbon dioxide.

Neutralization is relatively simple and reliable but typically requires automatic feed equipment, pH monitors/controllers, and multiple mixing tanks.

To reduce chemical use and costs, mixing of alkaline and acidic wastewaters should also be considered.

Chemical precipitation is another chemical treatment method often used to treat industrial wastewater. Chemical coagulation (rapid mixing) and flocculation (slow mixing) are used to precipitate dissolved wastewater contaminants and form floc particles, which settle readily in sedimentation basins. Chemical precipitation can effectively remove heavy metals and phosphorus from industrial wastewater. The major disadvantage of chemical precipitation is that it may generate relatively large amounts of inorganic sludge, which typically must be dewatered and landfilled. If the sludge contains metals at toxic levels or is otherwise hazardous, it must be disposed of as a hazardous waste. In addition, close operator attention and rigorous cleaning are necessary to maintain a mechanically reliable chemical feed system (U.S. EPA, 1980; Viessman and Hammer, 1985; and Weber, 1972).

Oxidation-reduction is used occasionally to remove pollutants from industrial wastes, for example, to reduce chromium from its hexavalent form to its trivalent form before chemical precipitation. Additionally, ozone oxidation may be used to remove dissolved organics and cyanide during pretreatment; however, alkaline chlorination of cyanide is a more common practice than ozone oxidation. Hydrogen peroxide or potassium permanganate may also be used for some industrial wastes. Oxidation-reduction systems have a high mechanical reliability. Offgases must meet air pollution requirements, however, and oxidation-reduction may not be economically attractive in some cases (Eckenfelder, 1982, and Weber, 1972).

In the ion exchange process, ions held by electrostatic forces to charged functional groups on a solid surface are exchanged for ions of similar charge in the wastewater. Ion exchange may be used for removing heavy metals, ammonia, and radioactive pollutants. The process is reliable and relatively easy to operate if automatic controls are used.

Ion-exchange systems require periodic monitoring, inspection, and maintenance, and pretreatment of the wastewater may be required to prevent resin fouling. Scaling can occur where wastewaters high in magnesium and/or calcium are treated. In addition, disposal of waste brine and rinse water is required. Recovery of valuable chemicals may be possible (Cherry, 1982; U.S. EPA, 1980; and Weber, 1972).

BIOLOGICAL PRETREATMENT. Biological pretreatment of industrial wastewater may be used to reduce BOD/suspended solids loads to WWTPs, degrade potentially toxic organic compounds, or reduce nutrient levels. Biological systems include activated sludge, lagoons, trickling filters, rotating biological contactors, and anaerobic processes. Where wastes are compatible, however, economies of scale often indicate that WWTP treatment of biodegradable wastewater better serves the community than does installation of several biological pretreatment systems.

The activated sludge process uses an aeration tank in which wastewater and microorganisms are mixed; the microbes biooxidize the waste matter and synthesize new cells, and the biological solids are then removed by final settling. Several modifications of the activated sludge process are available. The one selected should best meet the pretreatment requirements. The process typically is reliable, but sludge disposal, aerosol and odor potential, and energy consumption may cause problems. Skilled operators are required for optimum performance (Reynolds, 1982, and U.S. EPA, 1980).

Aerated lagoons are medium-depth (typically 2 to 4 m [6 to 12 ft]) basins and function similarly to the activated sludge process but without recycle. In addition to being reliable, aerated lagoons require only basic wastewater operator skills. Air emissions from the lagoons must meet air pollution requirements, however, and potential effect on groundwater from lagoon seepage must be evaluated in design and operations. A liner may be required (Metcalf and Eddy, 1991). Facultative lagoons are typically 1- to 2.5-m (3- to 8-ft) deep basins in which wastewater is stratified into an aerobic surface layer, a facultative layer, and an anaerobic bottom layer. Facultative lagoons are also reliable and require basic operator skills. Like aerated lagoons, air and groundwater discharges must be evaluated and appropriately addressed.

Trickling filters consist of a fixed bed of rock or plastic media over which wastewater is distributed for aerobic biological treatment. Biological slimes that form on the media assimilate and oxidize substances in the wastewater. The biomass repeatedly falls off the media (sloughing) and must be removed in a settling tank following the trickling filter. Though not as efficient as activated sludge systems, trickling filters are typically reliable. They have limited flexibility, however, are susceptible to upsets, and may have difficulty operating in cold weather (Metcalf and Eddy, 1991, and Reynolds, 1982).

Rotating biological contactors are fixed film reactors normally consisting of plastic media mounted on a horizontal shaft in the tank. As wastewater flows through the tank, the media, approximately 40% immersed, are slowly rotated. Biomass on the media assimilate (oxidize) the organics. Excess biomass is stripped off the media by rotational shear forces and is subsequently removed during final settling. Rotating biological contactors perform well, with reliability, unless organic loads are high or temperatures are below 13°C (55°F). Odor may be a problem, and sludge treatment and disposal is required. Additionally, facilities with large flows may incur high capital and operating costs.

A packed-bed reactor consists of a reactor that is packed with a medium to which the microorganisms can become attached. Wastewater enters the bottom of the reactor through an appropriate inlet chamber. Air or pure oxygen necessary for the process is introduced with the wastewater.

Anaerobic processes include contact, filters, fluidized-bed reactors, and lagoons. Anaerobic contact provides for separation and recirculation of seed microbes, thus allowing retention periods of 6 to 12 hours. The anaerobic filter promotes growth of the anaerobes on a packing bed and can be designed for upflow or downflow operation. For the fluidized bed reactor process, wastewater is pumped up through a sand or plastic bed, which supports microbial growth; effluent recycle is practiced. Anaerobic lagoons are commonly used to pretreat meat-packing and other high-strength organic wastewaters. Anaerobic processes typically are reliable, but odor problems and process upsets may occur (Eckenfelder, 1989). Preference depends on waste strength, temperature, wastewater chemistry, and other factors.

CROSS-MEDIA POLLUTANTS. In selecting pretreatment options, cross-media pollutant generation must be considered. Many pretreatment facilities generate sludge that requires handling, treatment, and disposal. The treatment and disposal of sludge, especially if it exhibits hazardous waste characteristics, can be expensive and cumbersome considering the multitude of sludge and hazardous waste regulations at the local, state, and national levels. Some pretreatment processes may result in air emissions, such as off-gases from air stripping of certain industrial wastewaters, that must comply with applicable air pollution standards. Processes such as ion exchange, ultrafiltration, and reverse osmosis result in reject streams requiring disposal.

SAFETY CONSIDERATIONS. Selection of pretreatment options also involves safety considerations. Under certain conditions, electrical and mechanical equipment, if incorrectly installed, inadequately maintained, or improperly used, can cause electrical shock or other bodily injury. Wastewater gases and pathogenic microorganisms can also create health hazards. Chemicals, such as chlorine, sulfides, or ammonia, that are either present in the wastewater or added during treatment, can create noxious vapors or otherwise cause acute or chronic injuries to plant personnel or the public if control measures are inadequate.

OFF-SITE PRETREATMENT. In formulating management strategies for the pretreatment of wastewater, alternatives for off-site pretreatment should also be considered. Off-site pretreatment generally means removing all or part of the wastewater generated by a facility and pretreating it at another location so that it is suitable for disposal. An off-site pretreatment facility may or may not be a Resource Conservation and Recovery Act (RCRA) hazardous waste treatment facility depending on whether the wastes accepted meet the RCRA definition of hazardous and whether the facility is RCRA permitted. Typically, the off-site facility is nearby and designed to treat specific types of wastewater from several firms in the area, such as plating shops or printed circuit facilities, at a lower cost than comparable treatment by the individual facilities generating the waste streams. The benefit of off-site treatment is that it can eliminate the need to install a costly pretreatment system for treating what may be a relatively small waste stream. However, off-site treatment facilities are typically only conveniently and economically available in certain metropolitan areas and may be subject to strict regulatory requirements that can result in the imposition of conditions and constraints on the wastewater-generating facility.

PROCESS MONITORING. The goal of any industrial wastewater pretreatment management strategy is to achieve cost-effective regulatory compliance by implementing waste minimization, wastewater recycling, water conservation, and wastewater treatment by the most appropriate treatment processes. To determine whether this goal is being achieved, the strategy must include a monitoring component that provides information on the effectiveness of the strategy and allows for necessary corrections. The primary purpose of monitoring is to ensure and verify that compliance with regulatory requirements, such as discharge permit conditions or categorical discharge standards, is consistently met.

However, the results from a properly designed discharge monitoring program (this includes adequate testing frequency, rapid analysis turnaround times, and analysis for the appropriate parameters) can also provide information on the efficiency of the pretreatment system and help improve the cost-effectiveness of the treatment system by offering process control data for reducing operating costs such as chemical and power usage rates.

Beside providing economic benefits, a good monitoring program is useful in assessing the effect of process or raw material changes or other waste minimization efforts, detecting potential upsets with the system that could cause discharge violations or slug discharges, and estimating loading surcharges that may be imposed by the WWTP receiving the discharge. The monitoring of the incoming wastewater also allows for improved process control, particularly in processes involving chemical addition or in the activated sludge process. Industrial facilities are increasingly using statistical process control techniques to ensure compliance.

Wastewater effluent monitoring can be performed by either the industry (self-monitoring) or the regulator with jurisdiction over the facility's discharge. Facilities with limited laboratory capabilities may enlist help from a contract laboratory for both sample collection and analysis. Industrial self-monitoring can be required by the regulator to meet reporting requirements for baseline monitoring reports or periodic reports on continued compliance required by federal regulations, as well as to ensure that the pretreatment system is operated in a compliant manner.

The results of any self-monitoring tests for regulatory purposes, whether performed in-house or by a certified laboratory, must be submitted to the regulator for review. Any violations may be subject to enforcement actions. Any self-monitoring samples collected for regulatory purposes in the U.S. must be analyzed in accordance with EPA-approved procedures. Informal self-monitoring is sometimes performed in house to facilitate the operation of the pretreatment system and check on its response to operational changes. This can be done using test kits or other rapid analytical means as long as the results obtained are accurate. Rapid self-monitoring through the use of test kits or on-site analytical instruments can provide a quick indication of pretreatment system upsets or problems. This feedback allows for the implementation of corrective measures, including the recycling or storage of a noncompliant effluent until the pretreatment system is operating properly again to prevent discharge violations.

Results from self-monitoring using nonapproved U.S. EPA methods do not have to be reported to the WWTP unless specifically requested by the WWTP, and they do not count toward determination of compliance. However, all analytical discharge or effluent results obtained using U.S. EPA-approved methods become part of the compliance history and must be reported. The facility is subject to enforcement if these results are in violation of the limits.

An effective monitoring system is integral to the pretreatment system design and should be planned at the conceptual stage. Inclusion of properly designed monitoring points and equipment as part of the total system design can significantly reduce future monitoring costs and provide for improved pretreatment system operation.

Compliance monitoring by a WWTP or other regulatory agency represents the second type of monitoring most industrial facilities face. This monitoring can either be scheduled or unannounced. By requiring an industrial facility to install appropriate monitoring or sampling points that are continuously accessible to the WWTP, the WWTP may set up sampling equipment or obtain grab samples on either short notice to the industry or totally unannounced.

Finally, in the event of spills, slug discharges, or chronic violations, a WWTP may initiate a "demand" monitoring program against a facility in response to the problem. Once the problem has been resolved and continuous compliance is achieved, the demand monitoring program is usually rescinded and a normal monitoring schedule is resumed.

COST ANALYSIS. In developing a management strategy for the control of industrial wastewater from a facility, it is important to develop a comprehensive cost analysis for the different options under consideration. Although cost figures strongly depend on local conditions and regulatory requirements, certain elements should be incorporated into the facility's management strategy before selecting a pretreatment option. These include determining the capital cost of the pretreatment system, the operating cost (including costs of chemicals, energy, labor, compliance, and residual disposal), and maintenance costs to keep the system operating in a manner that allows for continuous compliance.

Determining which pretreatment alternative to select should include a review of return on investment goals. The selection of the pretreatment alternative should be based on lowest life-cycle cost and highest return on investment. The performance and treatment capability of a pretreatment system will largely be driven by the regulatory requirements imposed on the discharge. In some instances, however, when the treatment of compatible pollutants is involved, it may be more cost-effective for the WWTP to treat the wastewater even though a higher user fee is charged than for the industry to pretreat its wastewater and discharge lower pollutant loads to the WWTP.

When developing a cost analysis, it is important to consider the many variables and factors in the pretreatment process, most of which will change over the lifetime of the system's operation (the life-cycle cost). An estimate of these changes, as well as the effect of any anticipated regulatory changes, residual disposal restrictions, or other applicable considerations should be incorporated into the cost analysis and, ultimately, the management strategy.

Cost information databases for pretreatment of industrial wastes are limited. This is because of the almost infinite variety of pretreatment waste streams to be treated. This variation makes it difficult to develop a large database of treatment costs. There are, however, sources of information on the more common waste streams that can be used as a starting point for the estimates, a check on order of magnitude costs, or a rough comparison of treatment options. In addition, sources of information usually exist as case studies for specific waste streams. However, these case studies typically are limited in the number of technologies compared. Industrial associations and trade groups, such as the American Petroleum Institute or the Chemical Manufacturers Association, are also a good source of information on specific waste streams because they may track and record these data for their members. It is important to use current and accurate cost and performance data because the technologies and treatment costs constantly change.

REFERENCES

Cherry, K.F. (1982) *Plating Waste Treatment.* Ann Arbor Science, Ann Arbor, Mich.

Eckenfelder, W.W. (1982) *Water Quality Engineering for Practicing Engineers.* Cahners Books, Boston, Mass.

Eckenfelder, W.W. (1989) *Industrial Water Pollution Control.* McGraw-Hill, Inc., New York, N.Y.

Metcalf & Eddy, Inc. (1991) *Wastewater Engineering Treatment, Disposal, and Reuse.* McGraw-Hill, Inc., New York, N.Y.

Reynolds, T.D. (1982) *Unit Operations and Processes in Environmental Engineering.* PWS-Kent, Boston, Mass.

U.S. Environmental Protection Agency (1980) *Innovative and Alternative Technology Assessment Manual.* Office of Water Program Operations, Washington, D.C.

U.S. Environmental Protection Agency (1985) *Guidance Manual for the Use of Production-Based Standards and the Combined Wastestream Formula.* Permits Div. and Indust. Technol. Div., Washington, D.C.

Viessman, W., and Hammer, M.J. (1985) *Water Supply and Pollution Control.* Harper and Row, New York, N.Y.

Weber, W.J. (1972) *Physicochemical Processes for Water Quality Control.* Wiley-Interscience, New York, N.Y.

S*UGGESTED READING*

U.S. Environmental Protection Agency (1991) *Pretreatment Facility Inspection.* 2nd Ed., Office of Water Enforcement and Permits, Washington, D.C.

Chapter 4
Flow Equalization

Flow equalization is the process by which operational parameters such as flow, pollutant levels, and temperature are made more uniform over a given time frame (normally 24 hours) to reduce the downstream effects of the parameters. The need for flow equalization is determined primarily by the potential effects the waste stream will have on the receiving body of water or treatment facility. This effect is determined by two key components: (1) the variability of the operating parameters to be equalized (including toxicity) and (2) the volume of the flow being discharged. Defining the need for flow equalization requires sufficient background information on these two factors, the relative cost of constructing and implementing effective flow equalization, and the cost savings by reducing the effects on downstream treatment systems.

The objectives of this chapter are to provide information on the various types of flow equalization processes used for pretreatment of industrial waste streams, consider the effects of each of these processes, and provide basic design criteria for each. However, because flow equalization is an integral part of the overall treatment system, all discussions in this chapter are related to and should be reviewed in context with other sections of this book. Therefore, justification for flow equalization should be a part of the management strategy discussed in Chapter 3 and the processes to deal with specific pollutants discussed in Chapters 6 to 10. For example, the need for flow equalization should be evaluated relative to other potential solutions, such as process or operation changes or other pretreatment.

The need for flow equalization as a pretreatment operation is generally determined by the ability to meet end-of-pipe permit limits when considering

- The variability of the pollutants contained in the stream and the resulting effects of this variability;
- The cost of providing an equalization system sufficient to eliminate excess variability; and
- The cost of alternative measures, for example, chemical or tertiary (postbiological) treatment systems such as granular activated carbon polishing.

Given the effluent criteria (by permit) and assuming a goal of the facility to achieve 100% compliance, the factors for designing and implementing an effective flow equalization program are determined by the variability of the pollutants and the downstream effects of this variability on the treatment facility or receiving waters. The direct benefits of flow equalization are observed in reducing the effects of controlled pollutants on the receiving facility or body of water. As a pretreatment to a biological facility, equalization may prevent potentially costly disruption of microbiological activity resulting from shock loads coming through the system or reduced efficiencies resulting from cyclical loading of biochemical oxygen demand (BOD) and/or toxins such as phenolic compounds and heavy metals. As part of a primary treatment facility, equalization can serve as the major unit operation before discharge by leveling the mass flow rates of all components.

Additional benefits are observed through savings on construction costs and treatment chemicals and improved process efficiencies. By eliminating wide variations in effluent stream parameters, the industrial discharger can experience improved operations and greater compliance with permit regulations.

Flow equalization processes

There are four basic flow equalization processes discussed in this chapter:

- Alternating flow diversion,
- Intermittent flow diversion,
- Completely mixed combined flow, and
- Completely mixed fixed flow.

The alternating flow diversion system, shown in Figure 4.1, is designed to collect the total flow of the effluent for a given period of time (normally

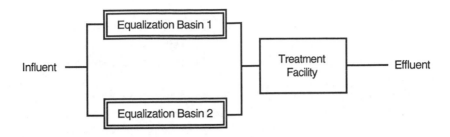

Figure 4.1   Alternating flow diversion equalization system.

24 hours) while a second basin is discharging. The basins alternate between filling and discharging for successive time periods. Thorough mixing is maintained so that the discharge maintains constant pollutant levels with relatively constant flow. This type of system provides a high degree of equalization for a given size basin by leveling all discharge parameters. The disadvantage to this type of system is the high construction cost associated with storing the volume of the waste stream for the time period used.

The intermittent flow diversion system, shown in Figure 4.2, is designed to allow any significant variance in stream parameters to be diverted to an equalization basin for short durations. The diverted flow is then bled back into the stream at a controlled rate. The rate at which the diverted flow is fed back into the main stream will depend on the volume and variance of the diverted water so as to reduce downstream effects.

The completely mixed combined flow system, shown in Figure 4.3, is designed to provide complete mixing of multiple flows combined at the front

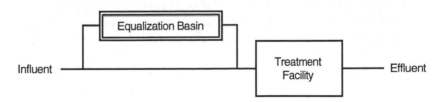

Figure 4.2   Intermittent flow diversion system.

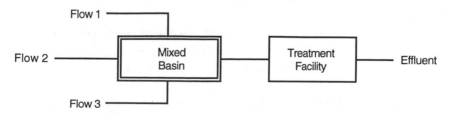

Figure 4.3   Completely mixed combined flow system.

end of the facility. This type of system can be used to reduce variance in each stream by thorough mixing with other flows. This system assumes that the flows are compatible and can be combined without creating additional problems.

The completely mixed fixed flow system, shown in Figure 4.4, is designed as a large completely mixed holding basin before the wastewater facility that levels variations of influent stream parameters and provides a constant discharge.

Each of these systems requires different design criteria. Therefore, the first step in the selection process is to define the type of variability the system will be equalizing. Then the appropriate criteria can be used for designing the facility.

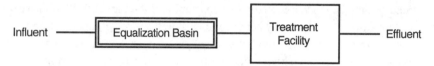

Figure 4.4   Completely mixed fixed flow system.

*D*ESIGN OF FACILITIES

The design of equalization facilities begins with a detailed study to characterize the nature of the wastewater and its variability. This study also should include gathering data on both flow and all pollutants of consequence.

The costs of the studies described (ranging up to $50 000 or higher) can be but a small fraction of the total construction cost but may represent a significant part of the total design cost. The budgetary allowances in the preliminary design stages must be made with the realization that the total cost of the construction may be reduced significantly if the design engineer can better characterize variability and, therefore, optimize the design by properly selecting the best type of equalization system.

A primary consideration in determining the data to be collected is the effect of the effluent on downstream facilities. The most significant quantity is the mass flow rate; therefore, data on both flow and concentration (in terms of BOD, total suspended solids [TSS], or other significant variables) must be measured on a time series basis. Previous studies have indicated that data of this type tend to be normally distributed so that the average mass flow from the sampled values may be taken as an estimate of the true average mass flow.

Determining the amount of data to be collected is one of the most difficult decisions in setting up a sampling program. The variables of interest are generally well known or easily ascertained, but the number of samples to be taken is difficult to evaluate. Because the data to be collected will be time series, the normal concepts of attempting to obtain random samples cannot be met. Time series data are by nature not random. Therefore, the program must contain a sufficient number of samples to allow proper characterization of the statistical parameters in question.

The following guidelines should be considered in determining the number of samples. For cyclical data, a minimum of two cycles must be collected. The spacing of data should be small enough to have a reasonable probability of measuring peak or minimum values. Where seasonal considerations are known to be important, at least one sampling program should be conducted during each season. The time constraints of a design project may not allow this sampling frequency in all cases. For the purpose of designing flow equalization, a minimum recommendation for industrial waste sampling is 2 weeks of data for variables of primary concern (that is, chemical constituents, chemical oxygen demand [COD]/BOD, or TSS). The samples should be collected every hour for the first day using an auto discrete sampler, and composite 24-hour samples should be collected for the remaining 13 days. Additionally, strip chart flow recording and/or real-time total organic carbon analyzers can also be used to determine variability. If possible, hourly flow data should be gathered for the entire 2-week sampling period.

The nature of the industrial processes is such that even if the effluent variability can be characterized properly at one point in time, the variability undoubtedly will change at some future time. After a sampling program has been completed, any anticipated changes in production, manufacturing techniques, or scheduling must be superimposed on the data so that the design is sufficiently flexible to meet anticipated operating conditions. These considerations along with factors for degrees of uncertainty constitute the management design criteria used by the design engineer.

For instance, McKeown and Gellman (1976) report that a time series examination of ten pulp and paper mills indicates the coefficient of variation (that is, the standard deviation divided by the mean) changes considerably from plant to plant. They verified that the effluent BOD and suspended solids follow a log normal distribution in most cases. They performed a time series analysis to study the variability of the raw waste and the variability generated by treatment and attempted to determine the relationship between effluent parameters, raw waste characteristics, and production characteristics. Results indicated that the variability of BOD and TSS for some of the mills studied are closely related, and there is a high seasonal occurrence at most facilities. The number of observations made of parameters such as BOD and TSS ranged from as low as 179 to as high as 333. This range indicates what might be considered an adequate number of samples to properly characterize variability.

While statistical analysis and extrapolation for confidence levels are an important part of determining the effects of variance (on reducing the potential effects), it is not the focus of this chapter. This chapter will assume that analysis has occurred and deal with the individual concepts.

ALTERNATING FLOW DIVERSION. Because the alternating flow diversion system is intended to hold the total flow for a fixed period of time (normally 24 hours), its design is based strictly on flow. Design criteria then depend on the variability of the flow, the standard deviation, and maximum flow for the given time frame.

Consider an industrial facility with a total daily flow and pollutant profile as indicated in Table 4.1. Assuming a 30-day period as representative of 1 month, the equalization basin can be designed using the data in Table 4.1 and a management design criterion of 110% of maximum flow, for example. Each of the equalization basins would be designed to hold 686.98 m^3. Therefore, the management design criterion (that is, 110% of maximum flow) becomes the dominant variable in this equation.

This variable is either given to the design company by plant management or assumed based on prior experience. To express this in a general equation, the following should apply:

$$V_t = D_c \, F_c \, T \, k \qquad (4.1)$$

Where

V_t	=	volume of each equalization basin, m^3;
T	=	time period for equalization;
D_c	=	management design criteria, %;
F_c	=	management flow criteria (f_a, average flow for given time period, or f_m, maximum flow for given time period), m^3/h; and
k	=	units conversion constant.

INTERMITTENT FLOW DIVERSION. Intermittent flow diversion systems are a bit more complex because of the need to consider the variance of the pollutants to be diverted, the average length of the variance, and the rate of discharge back to the system. Each of these factors has to be evaluated with respect to the effect on downstream processes, especially if biological systems follow. This type of equalization system is best used when variances are easily detectable, infrequent, and could have a dramatic effect on downstream processes. An example of this type of problem is phenol levels in an effluent stream.

The following steps should be applied to the design of the system.

- Step 1: determine the frequency and duration of variance to be diverted (this will allow design of the equalization basin).

Table 4.1   Industrial facility daily flow profile.

Day of month	Total flow, m³/d	Phenol levels, μg/L (ppb)
1	377.44	45
2	471.46	393
3	411.96	421
4	254.35	433
5	350.64	683
6	464.19	822
7	339.29	123
8	624.52	467
9	569.57	682
10	47.15	732
11	420.13	398
12	238.46	541
13	553.42	868
14	487.81	558
15	241.18	656
16	562.75	329
17	272.52	822
18	229.83	771
19	237.55	613
20	348.47	400
21	134.44	821
22	0.00	0
23	143.98	160
24	610.90	214
25	97.65	670
26	398.78	362
27	560.94	303
28	253.44	245
29	574.11	120
30	525.06	251
Average daily	360.59 (0.25 m³/min)	463
Minimum	0.00	0
Maximum	624.52	868

- Step 2: calculate the rate of controlled release of the diverted flow so as to maintain normal operation.
- Step 3: use the diverted volume to calculate the volume of the surge basin so that continuous flow to the treatment facility can be maintained.
- Step 4: verify that the equalized flow meets desired discharge limits.

As stated earlier in this section, data collection and system profiling are keys to effective design for this type of equalization system. An effective system is one that is automated based on electronic monitoring of the stream with diversion occurring as necessitated. Three examples of this technology are pH sensors to monitor for pH excursions, on-line gas chromatographs to monitor phenol excursions, and conductivity sensors to monitor total dissolved solids. Wide variations of these parameters can cause substantial damage to biological systems or receiving waters (especially if only primary treatment is being used).

From the example in Table 4.1, the phenol levels are observed to vary substantially from day to day because of variance in plant operation. Given the variability of the plant operation, it becomes necessary to divert flow at this facility to prevent permit violation and bleed the diverted flow back as the concentrations allow.

The phenol levels shown in Table 4.1 are 24-hour composite samples with a discharge limit of 500 ppb. Further analysis of individual samples indicated the problem was generated during two periods over the course of the day lasting approximately 3 hours each. They occurred between 3:00 and 6:00 p.m. and 11:00 p.m. and 2:00 a.m. Also observed during this time frame was an increase in flow rate to 0.473 m^3/min.

Therefore, the total volume to be diverted is

$$V_D = F_D\, T_D f_D\, k \qquad\qquad (4.2)$$

Where

$$
\begin{aligned}
V_D &= \text{volume of flow to be diverted per time period, m}^3; \\
F_D &= \text{flow rates diverted, m}^3\text{/min;} \\
T_D &= \text{time of diversion, hours;} \\
f_D &= \text{frequency of diversion, number/day; and} \\
k &= \text{conversion constant for unit, min/h.}
\end{aligned}
$$

Therefore,

$$
\begin{aligned}
V_D &= (0.473 \text{ m}^3\text{/min})(3 \text{ h})(2\text{/d})(60 \text{ min/h}) \\
V_D &= 170.28 \text{ m}^3\text{/d}
\end{aligned}
$$

The controlled discharge rate can then be established as

$$f_c = V_D/T \, k \qquad (4.3)$$

Where

f_c = controlled discharge rate, m³/min;
V_D = volume diverted, m³;
T = time period for return, hours; and
k = conversion constant for unit.

Therefore,

f_c = (170.28 m³/24 h)(1 h/60 min)
f_c = 0.118 m³/min

The volume of the surge basin can now be calculated. Recognize that 170.28 m³ of the total flow will be diverted and fed back to the stream at a constant rate. Therefore, the average flow for the remainder of the time is (360.09 – 170.28) = 189.81 m³ for the 18-hour period. This will equate to 0.1318 m³/min on a 24-hour basis. Correspondingly, to maintain this flow for the 6-hour diversion period, a surge basin equal in volume for the diversion time frame (6 hours in this case) at the average flow rate for the remaining period can be calculated as follows:

$$V_S = F_A \, T_D \, k \qquad (4.4)$$

Where

V_S = volume of surge tank basin, m³;
F_A = average flow rate without diversion flow, m³/min;
T_D = diversion time period, hours; and
k = unit conversion factor.

Therefore,

V_S = (0.175 6 m³/min)(6 h)(60 min/h)
V_S = 63.22 m³

As pointed out earlier, any excesses in design capacity are determined by management as part of design criteria and are not represented in the above calculation.

Combining the return of the diverted flow with the mainstream can be accomplished with in-line mixing or flash mixing just before downstream processes. The total combined flow (f_T) would be:

$$f_T = f_A + f_c \qquad (4.5)$$

$$= 0.118 \text{ m}^3/\text{min} + 0.176 \text{ m}^3/\text{min}$$

$$= 0.294 \text{ m}^3/\text{min}$$

Where

f_A	=	average flow rate without diversion, m^3/min, and
f_c	=	controlled discharge rate, m^3/min.

COMPLETELY MIXED COMBINED FLOW. The completely mixed combined flow equalization system is designed to address variability resulting from multiple flows coming from different sections of the plant that often generate impulse or step input changes to the wastewater treatment facility. The primary purpose of this system is to trim impulse variance or provide for a more gradual change in operating parameters.

Again, the volume of the equalization basin is determined by the effects the change in operating parameters will have on downstream systems. Because this is a more complex situation, design details will be approached from the simplest perspective: time and combined flows.

Therefore, the volume of the equalization basins V_e would be

$$V_e = (\Sigma f_i) \, T_e \, k \tag{4.6}$$

Where

f_i	=	individual flow rates, m^3/min;
T_e	=	time for equalization, hours; and
k	=	conversion factor for units.

If, for example, three flows come into the equalization basins with flow rates of 1.98, 0.567, and 0.189 m^3/min, respectively, and the desired equalization time is 1 hour, then the following would apply:

$$V_e = \Sigma \, (f_1 + f_2 + f_3) \, T_e \, k$$

$$= (1.98 + 0.567 + 0.189)(1 \text{ h})(60 \text{ min/h})$$

$$V_e = 164.16 \text{ m}^3$$

From here, the relative change in each operating parameter can be calculated using the formulas in the following section and converting the variability of the individual stream to variability in the total flow. This can be accomplished by the following formula:

$$\text{Var}_T = (\text{Var}_{pi}) \frac{f_i}{f_t} \tag{4.7}$$

Where

Var$_T$	=	variance in concentration of the total stream, ppm or ppb (mg/L or µg/L);
Var$_{Pi}$	=	variance in concentration of the individual stream, ppm or ppb (mg/L or µg/L);
f_i	=	flow of the individual stream, m^3/min; and
f_t	=	flow of the total stream, m^3/min.

For example, if the concentration of pollutant in an individual stream changes by 50 mg/L, the total stream would change.

$$\text{Var}_T = (50)\ (150/700) = 10.7\ \text{ppm}$$

This variance can be used in the calculation as the change in concentration of the combined and potential effect on the downstream system.

A typical industrial waste problem involves constant flow, with wastewater concentration as the only variable. The method described below may be used in the design of facilities to reduce this kind of concentration variability. The method assumes the data are approximately normally distributed.

Assume a completely mixed constant flow tank with a variable concentration input. If discrete samples are collected at uniform time intervals, Δt, the influent variance (s^2) may be estimated by

$$s^2 = [(C_i - C)^2]/(n - 1) \tag{4.8}$$

Where

C_i	=	influent concentration at the i time interval,
C	=	mean concentration, and
n	=	number of samples.

The influent coefficient of variation (v_o) is

$$v_o = s/C \tag{4.9}$$

An estimate of required equalization time based on the variation of the influent concentration and sampling intervals is

$$\theta = \Delta t/2[(v_o/v_t)/v_e]^2 \tag{4.10}$$

Where

θ	=	required equalization time, hours;
Δt	=	sampling interval, hours;
v_o	=	influent coefficient of variation for concentration, ppm (mg/L);

v_t =   average influent concentration, ppm (mg/L); and

v_e =   effluent variability coefficient, ppm (mg/L).

Both the influent and effluent coefficients of variation are based on discrete samples collected at uniform time intervals Δt.

Normally, the only information available will be raw wastewater characteristics, which provide v_o and Δt. The effluent variability v_e must be related to downstream requirements and, hence, is the primary design variable. It must be selected by the design engineer based on subsequent treatment units and effluent standards. Where the literature does not provide specific limits on acceptable variability, engineering judgment must be exercised.

The effluent variability V_e may be estimated as

$$V_e = \{[(C_e \max/C) - 1]/C\}/N \qquad (4.11)$$

Where

$(C_e)_{\max}$ =   the equalization tank effluent concentration not to be exceeded,

C =   mean value of concentration,

N =   cumulative standard normal for confidence level desired (confidence level is the probability that a specified concentration will not be exceeded).

Cumulative standard N may be selected from the abbreviated Table 4.2. Application of this method is illustrated in the example based on data presented in Table 4.3. Based on Table 4.3, $v_t = 698$ mg/L and $v_o = 158.6$ mg/L.

If downstream conditions (for example, the next treatment unit in line) restrict the effluent variability to 10% ($v_e = 0.1$), by using Equation 4.10 the required equalization time is

$$\theta = 1/2[(158.6/698)/0.1]^2$$

Specific restrictions on variability are uncommon, and a more realistic problem is to design an equalization tank so that the effluent does not exceed some specified value.

Table 4.2   **Selection of cumulative standard normal for desired confidence level.**

Confidence level	Cumulative standard normal N
90.0	1.282
95.0	1.645
99.0	2.327
99.9	3.091
99.99	3.719

Table 4.3 Hourly influent chemical oxygen demand (mg/L) data for 4-day period* (Wallace and Zellman, 1971).

Hour of day	First day	Second day	Third day	Fourth day
7	413	565	485	723
8	468	612	409	765
9	510	536	466	864
10	568	637	482	844
11	487	536	507	669
Noon	600	684	631	711
1	674	644	695	879
2	638	615	545	847
3	638	662	660	876
4	648	468	545	890
5	584	752	736	890
6	697	738	666	1 030
7	629	752	704	1 090
8	606	655	625	920
9	626	695	730	823
10	684	800	679	1 030
11	742	738	853	1 050
Midnight	729	380	612	1 010
1	884	708	504	736
2	638	678	606	882
3	677	648	599	812
4	1 210	608	5 651	832
5	995	738	590	867
6	780	662	631	775

* Average = 698 mg/L, maximum = 1 210 mg/L, and standard deviation = 158.6 mg/L.

These analyses can ultimately produce the type of curve shown in Figure 4.5. This type of graph allows subjective analysis of a particular tank size to determine how well it will suit requirements.

Assuming that a detention time of 3 hours tentatively has been selected based on the foregoing analysis and physical considerations at the plant in question, the effluent from this size tank would be expected to exceed the value of 800 mg/L approximately 5% of the time, or about eight samples per week. To reduce this expectation to fewer than two samples per week (that is, a confidence level of 99%) exceeding 800 mg/L, the detention time would have to be increased to approximately 7 hours. Similarly, if a confidence level of 90% (one sample in 10 or roughly two samples per day exceeding 800 mg/L) is acceptable, the size of the tank may be reduced to yield a detention time of approximately 2 hours.

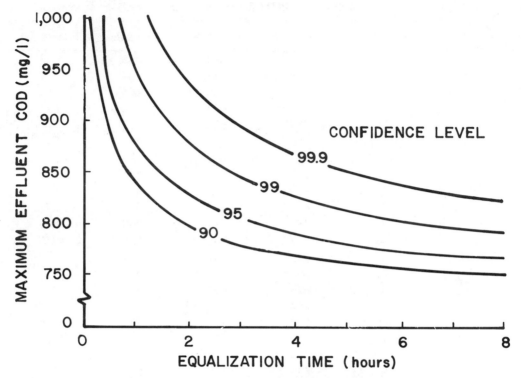

Figure 4.5 **Maximum effluent values as a function of equalization time and confidence level.**

CUMULATIVE FLOW CURVE. Equalization basins for individual facilities may be sized on the basis of a cumulative flow or mass diagram. The method is well known and has long been used for determining the storage required for water reservoirs. The graphic technique consists of plotting cumulative flow versus time for one complete cycle (24 hours for municipal facilities). Two parallel lines, with slopes representing the rate of pumping or flow from the equalization tank, are drawn tangent to the high and low points of the cumulative flow curve. The required tank size is the vertical distance between the two tangent lines.

The method is illustrated in Table 4.4 and Figure 4.6. The example shows hourly average flows from the Ewing Township, New Jersey, WWTP (La Grega and Heinick, 1978). The flow has accumulated starting at midnight and the cumulative flow is plotted as the curved line in Figure 4.6. The average flow for the day is represented by the line through the origin and the 24-hour cumulative flow value. This line represents the rate of constant flow from the equalization tank.

Table 4.4   Flow data for example problem.

Time	Flow rate,* m³/h	Cumulative flow, m³	Time	Flow rate,* m³/h	Cumulative flow, m³
Midnight	946	0	1	1 110	12 830
1	901	901	2	1 400	14 290
2	799	1 700	3	1 310	15 600
3	753	2 453	4	1 490	17 090
4	738	3 191	5	1 350	18 440
5	719	3 910	6	1 100	19 540
6	749	4 659	7	1 370	20 910
7	780	5 439	8	1 420	22 330
8	1 000	6 439	9	1 370	23 700
9	1 370	7 809	10	1 100	24 800
10	1 280	9 089	11	1 270	26 070
11	1 230	10 320	Midnight	1 230	27 300
Noon	1 400	11 720			

* Source of data: Flow from Ewing Township, New Jersey, WWTP.

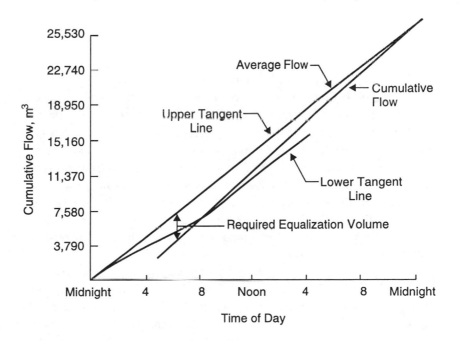

Figure 4.6   Cumulative flow curve.

The above procedure provides the tank size for the flow-time trace of a specific day. The variability and, thus, the amount of equalization required changes from day to day. Therefore, care is required in selecting a day or flow rate that is representative of the flow conditions to be equalized.

OPERATIONAL CONSIDERATIONS

MIXING REQUIREMENTS. Mixing of an equalizing vessel is a necessity. Typically, continuous mechanical mixing is best, although inlet arrangements may be such that the necessary homogeneity of soluble waste constituents is attained. If settleable and floatable solids are present, the wastewaters will need to be mixed to maintain a reasonably constant effluent concentration and prevent troublesome accumulations. For biodegradable wastes, the equalization tank will develop odor problems unless means of aeration are provided. The aeration and mixing systems may be combined (for example, floating surface aerators).

Although mixing power levels vary with basin geometry, 0.3 L/m$^3 \cdot$s (18 cfm/1 000 cu ft) of basin volume is considered the minimum to keep light solids in suspension (approximately 0.02 kW/m^3 [0.1 hp/1 000 gal]). Heavy solids like grit, swarf from machining, fly ash, and carbon slurries require considerably more mixing energy. The most common and practical approaches to mixing are baffling, mechanical agitation, aeration, or a combination of all three.

BAFFLING. Although not a true form of mixing and less efficient than the other methods, baffling prevents short-circuiting and, if effective, is the most economical. Over-and-under or around-the-end baffles may be used. Over-and-under baffles are preferable in wide equalization tanks because they provide more efficient horizontal and vertical distribution. The influent should be introduced at the bottom of the tank so that the entrance velocity prevents suspended solids in the wastewater from sinking to and remaining on the bottom. Additionally, a drainage valve should be present on the influent side of the tank to allow drainage of the tank when necessary. Normally, baffling is not advisable as a proper means of mixing wastewaters that have high concentrations of settleable solids.

MECHANICAL MIXING. Because of its high efficiency, mechanical mixing typically is recommended for smaller equalization tanks, wastewater with higher suspended solid concentrations, and waste streams with rapid waste strength fluctuations. Mechanical mixers are selected on the basis of laboratory pilot plant tests or data provided by manufacturers. If pilot plant results are used, geometrical similarity should be preserved and the power input per unit volume should be maintained. Because power is wasted in changing water levels by vortex formation, creation of a vortex should be reduced by mounting the mixer off center or at a vertical angle or by extending baffles out from the wall.

AERATION. Mixing by aeration is the most energy intensive of the methods. In addition to mixing, aeration provides chemical oxidation of reducing compounds, as well as physical stripping of volatile chemical compounds. It should be noted that some states will require an air discharge permit for volatile organic compound emissions to the atmosphere or classification of the equalization tank as a process tank.

Waste gases may be used for mixing if no harmful substance is added to the wastewater. Flue gases containing large quantities of carbon dioxide may be used to mix and neutralize high-pH wastewater.

DRAINING AND CLEANING. Equalizing systems should be sloped to drains and a water supply provided for flushing without hoses; otherwise, remnants after draining may cause odor and health nuisances.

REFERENCES

La Grega, M.D., and Heinick, R.M. (1978) Application of Equalization to Industrial Wastewaters. Paper presented at meeting of Water Pollut. Control Assoc. Pa.

McKeown, J.J., and Gellman, I. (1976) Characterizing Effluent Variability from Paper Industry Wastewater Treatment Processes Employing Biological Oxidation. *Prog. Water Technol.*, **8**, 147.

Wallace, A.T., and Zellman, D.M. (1971) Characterization of Time Varying Organic Loads. *J. Sanit. Eng. Div., Proc. Am. Soc. Civ. Eng.*, **97**, 257.

SUGGESTED READINGS

Downing, A.L. (1976) Variability of the Quality of Effluent from Wastewater Treatment Processes and its Control. *Prog. Water Technol.*, **8**, 189.

Fair, G.M., *et al.* (1963) *Water and Wastewater Engineering.* Vol. 1, John Wiley & Sons, New York, N.Y.

Foess, G.W., *et al.* (1977) Evaluation of in-line and side-line flow equalization systems. *J. Water Pollut. Control Fed.*, **49**, 1, 120.

La Grega, M.D., and Keenan, J.D. (1974) Effects of equalizing wastewater flows. *J. Water Pollut. Control Fed.*, **46**, 1, 123.

Novotny, V. (1987) Response of First Order Biological Treatment Processes to Time Variable Inputs. *Prog. Water Technol.*, **8**, 15.

Novotny, V., and Englande, A.J. (1974) Equalization Design Techniques for Conservative Substances in Wastewater Treatment Systems. *Water Res.* (G.B.), **8**, 323.

Smith, R., *et al.* (1973) *Design and Simulation of Equalization Basins.* ORM Rep. 670/2-73-046, U.S. EPA, Washington, D.C.

Chapter 5
Solids Separation and Handling

Although several unit operations in municipal wastewater treatment plants (WWTP) are designed to remove and concentrate suspended solids, it may be desirable or necessary to remove such solids before discharge to WWTPs.

Solids may be present in industrial wastewater in such quantities that they would interfere with proper operation of the WWTP.

In excessive concentrations, for example, suspended solids can affect both unit operations and collection systems. Unit operations may be adversely affected by organic solids (especially organic, putrescible solids) that are removed with grit in municipal systems, inorganic precipitates that can cause wear in pumps, and suspended solids that form relatively large, sticky masses that can clog treatment systems. In addition, introducing high concentrations of suspended solids to municipal systems can overload clarifier mechanisms.

In collection systems, high concentrations of readily settleable material can clog lines and pumping stations. Similarly, floating suspended solids can accumulate in pumping station wet wells. Besides being difficult and expensive to remove at this stage, these materials can cause objectionable odors if they are biologically degradable. Finally, pretreatment sometimes is required to remove a particular substance prohibited from the municipal system.

Suspended material in wastewater is generally classified according to its size and the technique used for removal:

- Large solids—objects at least 25 mm (1 in.) in diameter that will interfere with downstream flow and unit treatment operations.
- Grit—suspended matter with a settling velocity substantially greater than that of organics found in domestic wastewater. Grit consists of sand, gravel, and other dense materials.
- Settleable solids—those materials that settle from wastewater in the standard Imhoff cone test. Settleable solids are mainly particles with diameters greater than approximately 10^{-3} mm (1 μm); they usually do not include gritlike materials or particles larger in diameter than 25 mm (1 in.).   *[handwritten: micrometer or micron]*
- Colloids—substances made up of particles measuring 10^{-6} mm to 10^{-3} mm (0.001 μm to 1 μm) in diameter that will not settle unaided. These particles have surface charges that must be neutralized to allow particle agglomeration, flocculation, and settling. Chemical coagulation with an inorganic metal salt or synthetic polyelectrolyte is commonly used to remove these materials.   *[handwritten: nanometer]*

Suspended solids in industrial wastewaters may be either organic or inorganic. Although grit contained in these wastewaters enters industrial sewers predominantly with stormwater runoff, it also may come from washing operations in the pulp and paper, timber products-processing, food-processing, and similar industries. Mill scale from steel pickling operations has the characteristics of grit (inorganic composition and high settling velocities) and is ordinarily removed in treatment processes similar to those used for removing grit in municipal WWTPs. Settleable solids and colloidal materials also may be either organic or inorganic, depending on the type of industrial processing operation from which they originate. Moreover, dispersing agents such as surfactants, if used in industrial processes, may cause a high degree of stabilization of suspended solids, making their removal more difficult. Such situations must be approached on a case-by-case basis.

SUSPENDED SOLIDS CLASSIFICATIONS

Solids contained in water and wastewater are referred to as residue (APHA, 1992). Total residue (or total solids) is the material left after evaporation of the sample and drying of the residue in an oven at a defined temperature. Total residue includes both filterable residue (that fraction of the total residue retained by a filter) and nonfilterable residue (the part that passes through the filter). Suspended solids are filterable residue; dissolved solids are nonfilterable residue.

Total suspended solids (TSS; filterable residue) are determined by filtering a sample through a defined filter medium and determining the weight of the residue following drying in an oven to a constant weight. The drying temperature is typically 103 to 105°C (217 to 221°F), although temperatures as high as 180°C (356°F) may be used in certain instances. Samples dried at 103 to 105°C retain water of crystallization and some mechanically occluded water. As a rule, however, loss of carbon dioxide and of volatile organic materials is slight at this temperature. Samples dried at 180°C lose almost all of the mechanically occluded water present, but some water of crystallization may remain in the residue. The loss of carbon dioxide in organic matter increases as the drying temperature is increased.

The quantity of TSS measured on a particular sample is highly dependent on the filter paper used. Glass fiber filter paper has been the standard for measurement of TSS for several years. In addition, the quantity of suspended solids measured for a given sample depends on the physical nature of the suspended material, the thickness of the filter mat, and the amount and physical state of material deposited on the filter (occlusion).

Total suspended solids comprise "fixed" and "volatile" fractions. These fractions are determined by filtering the sample through a 22-mm (0.9-in.) disk until 200 mg of residue are collected on the filter. The residue is dried and weighed, and then ignited at a temperature of 550°C (1 055°F). The weight of the residue following ignition is the fixed suspended solids fraction. The difference between the fixed and total suspended solids weights is the volatile suspended solids fraction.

Because of the high viscosity of sludge, the TSS and volatile suspended solids contents of these are frequently ascertained by first determining the total solids (total residue) or total volatile solids (total volatile residue), respectively, for a particular sample. Total solids and total volatile solids includes both suspended and dissolved solids. The values must be corrected by subtracting the dissolved solids concentration contained in the sample.

REMOVAL METHODS—MAINSTREAM TREATMENT

The selection of methods for suspended solids removal depends on the following:

- Initial concentration of solids in the wastewater;
- Final concentration desired; and
- Size, settleability, thickening characteristics, and discrete or flocculent nature of the particles.

Techniques customarily used to remove suspended material from the main processing stream or from streams having TSS concentrations below 1% (10 000 mg/L) may be classified according to the following scheme. For each removal mechanism, the principal unit processes used in pretreatment operations also are listed:

- Removal by straining
  — Coarse screens.
  — Fine screens.
- Gravity separation
  — Grit removal by sedimentation.
  — Plain sedimentation (sedimentation unaided by coagulants or flocculents).
  — Chemical coagulation followed by sedimentation.
  — Flotation (either dissolved air flotation or induced air flotation, usually accompanied by the use of coagulants and flocculent aids).
- Filtration
  — Granular media filtration.
  — Precoat filtration.

Frequent pilot testing of the specific wastewater stream is necessary to determine its solids characteristics and compatibility with a particular process. Detailed designs for removal processes may be found in existing design manuals (Besselievre and Schwartz, 1976; Industrial Pollution, 1971; Metcalf & Eddy, 1991; Nemerow, 1978; Nemerow, 1991; U.S. EPA, 1975; Viessman and Hammer, 1985; and WEF, 1992).

STRAINING. Coarse or fine screens are used to strain solids from a waste stream. Screens with openings of 3 mm (0.118 in.) or larger are usually classified as coarse screens; those with openings less than 3 mm are classified as fine. The opening size required for a specific application is determined by the purpose of the screen, the unit operation to follow the screen, and the particle size that the screen can effectively and economically remove (WEF, 1992).

Coarse Screens. The most commonly used coarse screens are bar screens, which are primarily used to protect downstream equipment against damage or reduced efficiency from large floating solids. A bar screen is mounted in a channel at an angle of 10 to 90 deg to the flow. The horizontal acute angle is downstream. Design considerations include dimensions of the channel, the clear spacing between bars, depth of flow in the channel, method of cleaning, and control mechanism. Channel design should provide velocities of between 0.3 and 0.9 m/s (1 and 3 ft/sec) to avoid excessive sedimentation in the channel while preventing solids from being forced through the bars at higher velocities. The bars are typically equally spaced, 25 to 50 mm (1 to 2 in.) apart.

Head loss through screens varies with the quantity and nature of screenings allowed to accumulate between cleanings. Design values should range between 0.2 and 0.8 m (0.7 and 2.6 ft) for clean to partially clogged screens. Head loss created by a clean screen may be calculated by considering the flow and the effective area of screen openings (that is, the sum of vertical projections of the screen openings).

Manual cleaning may be cost-effective in a small system; however, infrequent or improper cleaning schedules may result in surges of high velocity due to plugging of the bar openings that decrease the screen's effectiveness and cause overflows in the channel. Mechanical cleaning is accomplished by rakes on loop chains or cables; the screens may be cleaned either from front or back (Figure 5.1). The rakes, powered by electric motors protected by overload devices, move over and between the bars, pulling the captured debris to a platform on top of the structure. Some screens are equipped with curved bars that are cleaned by a revolving rake. Mechanical cleaning reduces labor costs, provides more constant flow conditions, allows better screening capture, and causes less nuisance. Mechanical bar screens are the most common types of units in use. A parallel, off-line channel using a manual bar screen is sometimes constructed to allow continuous operation while the mechanical screen is being serviced.

Fine Screens. The rotary drum screen, tangential screen, and vibratory screen are the most commonly used fine screens for removal of fine nonflocculent and noncolloidal particles.

ROTARY DRUM SCREENS. The rotary drum screen, which is mounted in a channel and operates partially submerged, has a rotating cylindrical screen

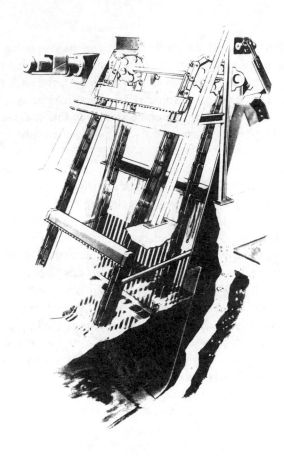

Figure 5.1   Catenary bar screen (WEF, 1992).

that revolves around the horizontal axis (Figure 5.2). As a rule, the liquid enters the center of the drum and flows radially through the screen; solids are
deposited on the screening fabric. In some cases, the liquid flows from the
drum perimeter to the center.

Solids left on the screen interior are elevated to the top of the drum where
they are deposited in a hopper for further dewatering and disposal and for
byproduct recovery. If the flow is from the exterior, the solids are retained on
the outer surface of the drum and removed by a scraper blade. Water jets
clean the screen to prevent blinding and clogging. Cleaning may be continuous or intermittent. The backwash may be actuated by increased differential
pressure (head loss) or conductivity. One of the advantages of a rotary drum
is low head loss or operating power requirement. Head loss across the screen,
including inlet and outlet structures, varies from 300 to 480 mm (12 to
19 in.). Head loss through the screen itself should be no more than 150 mm
(6 in.).

The screens are frequently constructed of stainless steel, manganese
bronze, nylon polyester, or alloy wire cloth. Openings commonly vary from

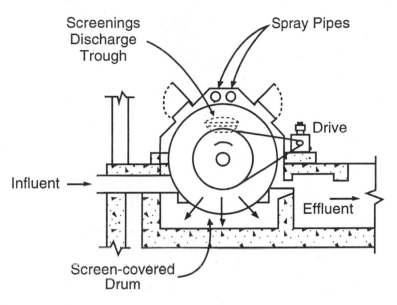

Screenings
Discharge
Trough

Spray Pipes

Drive

Influent →

Effluent →

Screen-covered
Drum

Figure 5.2   Rotary drum screen.

0.02 to 3 mm (0.000 8 to 0.1 in.). Opening size does not account for overall solids removal. The mat of removed solids provides a mechanism for removing smaller particles. The drums are from 0.9 to 1.5 m (3 to 5 ft) in diameter, 1 to 4 m (3 to 13 ft) in length, and rotate at approximately 4 rpm.

Rotary-type screens equipped with a screen or fabric with apertures of between 0.01 and 0.06 μm (3.9×10^{-4} and 2.4×10^{-3} mil) are also used for microstraining. This type of screening typically is used for a final polishing step.

TANGENTIAL SCREENS. Tangential, or static, screens are a second type of fine screen used for solids removal (Figure 5.3). The tangential screen has a curved surface consisting of a series of spaced wedge wire bars set at a right angle to the direction of flow and typically spaced about 1.5 mm (0.06 in.) apart. Above the screen and running across its width is a headbox with an inlet; a baffle may be used to reduce the turbulence of flow onto the screen. Oversize material travels along the screen surface and is collected and discharged at the bottom in various ways (for example, in a trough with a screw conveyor). The liquid passes through the screen and is discharged through a separate outlet. In addition to being self-cleaning, a tangential screen has two other advantages: incidental aeration as the process water is separated and the absence of moving parts. The screen may require a daily washing of approximately 5 minutes with steam or hot water to prevent blinding with grease. The unit normally is of stainless steel but may have a fiber glass housing.

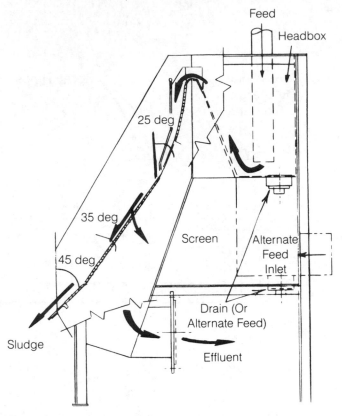

Figure 5.3   Inclined self-cleaning static screen (WEF, 1992).

VIBRATORY SCREENS. Vibratory screens are helpful in industries that use a lot of water. For example, in the food-processing industry, dewatering of solids is desirable either for disposal or byproduct recovery.

There are two basic types of vibratory screens: the circular center-feed unit and the rectangular end-feed type. In the center-feed type, solids are discharged in a spiral toward the center or the periphery (Figure 5.4). In the rectangular, end-feed type, solids are discharged along the screen toward the lower end.

GRAVITY SEPARATION. Suspended solids may also be removed by gravity, relying on the natural tendency of solid particles to settle or rise in a quiescent condition. Those with a specific gravity higher than the liquid settle; those having a lower specific gravity float.

Grit Removal. Grit is predominantly nonputrescible solids with a settling velocity greater than that of putrescible solids. Grit removal is important to protect downstream pretreatment equipment; prevent accumulation of heavy material in equalization, neutralization, and aeration tanks; and prevent accumulation in sewer pipes. Consideration should be given to locating grit-removal equipment close to the source of the grit. This can facilitate potential

Figure 5.4　Circular vibratory screen.

recovery and reuse and help prevent on-site sewer plugging. Two methods of removal are velocity control and aeration.

In velocity-controlled grit removal, a control section in the downstream channel provides a nearly constant velocity over a range of flows to vary the depth of flow in the channel as the volume changes. Control devices such as proportional and Sutro weirs are located 150 to 300 mm (6 to 12 in.) above the grit channel invert to provide grit storage and prevent resuspension of settled particles.

A velocity of approximately 0.3 m/s (1 ft/sec) allows heavier grit to settle out but carries through most of the lighter organic particles and tends to resuspend any that settle. Cleaning typically is mechanically controlled by adjustable time clocks. Continuously moving equipment is not recommended because of the abrasive nature of the grit.

Another type of controlled-velocity grit chamber that is increasingly popular in industry is the vortex-type grit chamber. In this unit, a vortex is generated hydraulically when the incoming flow is introduced tangentially near the top of the unit. Grit is literally "spun out" of the wastewater to the bottom of the unit where it is removed and dewatered by a conveyor-type device. Lighter organic solids remain in the liquid stream and are carried downstream to the next unit process. This unit has several advantages over other grit-removal devices. First, it has no mechanical parts in continuous contact with abrasive grit particles. Second, the system can be designed to remove even fine grit particles. The major disadvantage of the vortex grit chamber is the hydraulic head required for it to function. Generally, pumping to or from the unit is required. Also, head loss increases as the desired grit removal (the fineness of the grit particles) increases.

Grit removal is a significant issue in the fruit and vegetable canning industry. Field dirt washed off fruits and vegetables before processing can contribute large amounts of grit and dirt to the wastewater stream. Also, the quantities are highly variable, depending on weather conditions before picking. Rainy weather produces large amounts of grit.

A dragout tank is sometimes used by industries that handle large grit quantities. The iron and steel industry in particular typically handles mill-scale (which is high in both grit and oil and grease) in dragout tanks. Shown in Figure 5.5, the dragout tank is similar to a conventional rectangular sedimentation tank with a chain and flight collector mechanism. However, the solids are conveyed continuously up a sloped section out of the tank and into a hopper, avoiding the need to pump large quantities of grit.

Diffused air also may be used in a grit chamber for grit removal. The heavy particles settle out and the lighter organic particles are suspended by the air and carried out. Recommended air rates are 5 to 12 L/s per linear meter of tank (3 to 8 cu ft/min/ft) with provisions to vary the air flow. The higher rates should be used in tanks of larger cross sections. Detention times for effective removal range from 1 to 3 minutes at maximum flow rates. The inlet and outlet structures of the grit chamber should be designed to prevent short-circuiting. The influent should be introduced directly into the

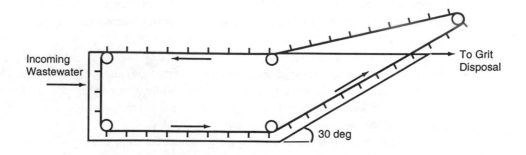

Incoming Wastewater

To Grit Disposal

30 deg

Figure 5.5 Typical grit dragout tank.

Pretreatment of Industrial Wastes

circulation pattern established by the air diffusion, and the outlet should be at right angles to the inlet. Dead spaces can be avoided by proper geometrical design of grit-collecting and air diffusion equipment. Mechanical cleaning is recommended.

The size and specific gravity of grit particles, as well as temperature of the wastewater, affect the rate of solids settling. Historically, most grit removal designs have been based on a grit specific gravity of 2.65 and a wastewater temperature of 16°C (61°F). However, the removal of finer grit is proving to be attainable and is desirable to preserve downstream equipment and processes. Both aerated and velocity-controlled grit chambers are designed for peak flow to provide more uniform grit removal over a wider flow range.

Plain Sedimentation. Plain sedimentation entails holding the liquid in a quiescent, or controlled, low-velocity stage long enough for solids to settle. Solids subject to removal by sedimentation are of specific gravity less than grit and, therefore, require a longer time for settling. Design parameters include surface area of the sedimentation tank, detention time, tank depth, surface overflow rate, and weir overflow rate. Peak flow rates typically are used in sizing sedimentation tanks.

The surface loading rate is the rate of wastewater flow over the surface area of the sedimentation unit. Typical values are approximately 80 to 120 m^3/m^2·d (2 000 to 3 000 gpd/sq ft) at peak flow rates. When the surface area of the tank has been established, the detention time can be determined by the tank depth. Primary sedimentation tanks generally provide detention times of 90 to 150 minutes at average flow rates. Tank depths vary from 2 to 5 m (7 to 16 ft), 4 m (13 ft) being most commonly used. Sufficient depth is necessary to prevent scour along basin bottoms and for solids storage; however, excessive retention of solids may result in an anaerobic condition.

Recommended weir loading is approximately 124 m^3/d per linear meter (10 000 gpd/ft), at average flow. Tanks may be circular or rectangular.

Circular tanks may be center fed or peripherally fed. In a center-fed clarifier (Figure 5.6), the influent enters a circular well that distributes the flow equally in all directions. The cleaning mechanism (or sludge scraper with two or four arms) is supported from a center shaft and revolves slowly. Scum removal blades may also be supported by the arms.

The peripherally fed design incorporates a suspended circular baffle a short distance from the tank wall. The influent is discharged tangentially at the base and flows spirally around the tank. The clarified liquid flows over a central weir; scum and grease are confined to the surface of the annular space.

Rectangular sedimentation tanks often are used where site space is limited. However, unless properly baffled, they are more subject to short-circuiting than are circular tanks. At excessive surface overflow rates, solids carryover can occur. Rectangular tanks should be designed with influent channels across the inlet end and effluent channels at the outlet. Some

Figure 5.6   Center-fed circular clarifier.

consideration should be given to inlet, effluent, and/or midtank baffling. Sludge-removal equipment may be a pair of looped conveyors or chains to which fiber glass or wooden flights are attached (Figure 5.7). Many plants have changed from redwood flights and iron chains to fiber glass to speed replacement or maintenance of the drive components. The flight and chain apparatus moves along the tank bottom at slow speeds, between 0.01 and 0.02 m/s (2 and 4 ft/min), scraping the settled solids to hoppers. In addition, the returning flights move scum to the end of the tank for collection.

An alternative cleaning mechanism is a bridge traveling up and down the tank on rails supported on the sidewalls (Figure 5.8). The bridge has one or more scraper blades that are lifted above the sludge on return travel. Scum also may be moved by water sprays or scrapers attached to the bridge.

Chemical Coagulation. Gravity separation can be enhanced with coagulants and coagulant aids, which are added to wastewater to promote flocculation. Coagulants are simple, water-soluble electrolytes (inorganic salts), inorganic acids, and bases. Iron, aluminum, and calcium salts are the most effective

Figure 5.7 Rectangular clarifier with wood flights mounted on parallel chains.

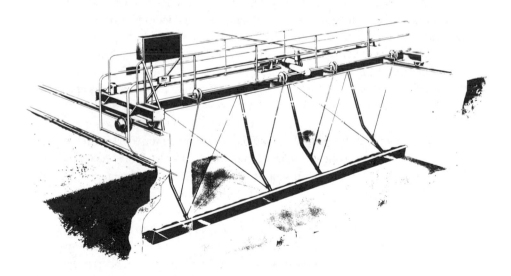

Figure 5.8 Rectangular clarifier with traveling bridge sludge collector.

coagulants. Specific coagulants regularly used are lime ($Ca(OH)_2$), alum ($Al_2(SO_4)_3$), ferric chloride ($FeCl_3$), ferrous sulfate ($Fe_2(SO_4)_3$), and sodium aluminate ($NaAlO_2$). Beside these conventional coagulants, certain other

---

chemicals have found favor in industry. In the meat-processing industry, for example, sodium lignosulfonate and calcium lignosulfonate are used when protein recovery is desired. These salts, byproducts of pulp and paper manufacturing, have been found to precipitate soluble protein for recovery at pH 3.5 to 4.0 in meat-packing plants and in seafood-processing plants. Beside protein recovery, soluble biochemical oxygen demand (BOD) removals of 70 to 90% have been observed (Hopwood, 1980; Sherman, 1979).

Other naturally derived products are also used to recover protein in the food industry. Chitin (a byproduct of shrimp and crab processing) and carrageen (an extract of seaweed) are used as coagulant aids where metal salts or polymers would affect the quality and value of the recovered product.

Coagulant aids include long-chain organic molecules with characteristics of polymers and electrolytes. Polyelectrolytes, bentonite, and activated silica are other commonly used coagulant aids.

An effective method for selecting a coagulant or coagulant aid and determining optimum dosage and pH is a standard jar test in conjunction with measurement of the zeta potential of the wastewater (Weber, 1972). Jar testing is particularly important with industrial wastes: Wastewater characteristics vary substantially between processes and even between functional units of the same process. Also, waste characteristics change throughout the day because of batch processing and cleanup activities. In the dairy industry, for example, pH fluctuations of up to 10 units are not uncommon when switching from processing to cleanup. Thus, jar testing should be performed on representative equalized samples where substantial variation is expected.

Components of a chemical addition system include facilities for chemical storage, chemical feeding, chemical mixing, rapid mixing of chemicals and wastewater, and flocculation.

Chemical storage facilities vary depending on the chemical chosen, the state of the chemical (liquid or dry), and the size of pretreatment facilities. Most coagulants can be bulk purchased in railroad car or truckload lots, although they are also available in smaller quantities. The cost advantages from bulk purchase should be weighed against construction costs and potential chemical deterioration over time.

Chemical feed systems are designed for both dry and liquid feed. Coagulants, in solid form, are typically converted to solution or slurry before introduction to the wastewater. Some coagulants, such as alum, are noncorrosive in the dry state, but become corrosive in liquid form; therefore, equipment for handling the liquid must be designed to resist corrosion.

A dry feed installation consists of a hopper, feeder, and dissolving tank. The design must take into account the particular chemical's characteristics and minimum and maximum wastewater flows. Some chemicals require vibration or agitation to prevent bridging and promote continuous flow. The dry feeder may be either volumetric or gravimetric, the latter being more accurate. Polymers are difficult to dissolve, and because of their diversity, no one scheme for dissolving and feeding suits all applications. Suppliers offer recommendations on dissolving and feeding of their polymers. A widely used system is a dissolving tank with a mixer and metering pump. Provision is often made for wetting the powder before introduction to the tank to prevent formation of "fish eyes," which lengthen mixing time and can reduce effectiveness of the polymer, requiring more of it to be added.

Liquid feed systems use a pump or rotating dipper for chemicals that are more stable and more readily fed as liquids; in situations where handling a dust or relatively dangerous chemical is undesirable; or when the chemical is available only in liquid form. Piston, positive displacement diaphragm, and balanced diaphragm pumps are frequently used. Chemical feeder controls may be manual, automatically proportioned to flow, or a combination of both.

The rapid dispersal of the chemical reagent throughout the waste stream is of primary importance to allow sufficient agitation and mix time. Typically, mix tanks perform this function, although pump suction and discharge lines also have been used successfully. Rapid mixing is facilitated by a mixer, normally an impeller or propeller mixer mounted at an off-center angle. In-line mixer baffling also has been used with success. Following rapid mix, an area for slow mixing should be provided to permit formation of aggregates or flocs from the finely divided matter. Detention time for flocculation should be 20 to 30 minutes at design flow.

The energy required to generate the needed velocity gradient for flocculation can be applied by hydraulic, air, or mechanical means. Mechanical agitation is preferred because it produces a more uniform energy distribution so that delicate flocs will not be sheared at the full range of wastewater flows. Efficiency and adaptability of mechanical flocculation allow modification to existing sites.

Flotation. Flotation is a unit process in which light suspended solids or liquid particles are separated from the liquid phase. (See Chapter 6 for a discussion of flotation to remove oil.) Factors to consider during design include operating pressure, air-to-solids ratio, float detention time, surface hydraulic loading, and percent of recycle flow. In addition, pilot testing is recommended to develop reliable design criteria.

Two conventional methods of flotation are gravity flotation and dissolved air flotation (DAF). Gravity flotation relies on the natural tendency of light suspended solids, often referred to as scum, to float to the liquid surface. Such common phenomena are used in conjunction with sedimentation. Dissolved air flotation introduces fine air bubbles into the liquid. The bubbles, which attach themselves to, or entrap themselves within, the particles, float to the surface to form a layer that can be skimmed. The components of a DAF unit include a pressurizing pump, retention tank, pressure-reducing valve, air injection equipment, and flotation tank (Figure 5.9).

Figure 5.9   Dissolved air flotation unit.

There are two types of dissolved air flotation. For one type, the wastewater flow is pressurized with air at 100 to 300 kPa (1 to 3 atm), held typically for up to 3 minutes in a retention tank, and released to atmospheric pressure in a flotation chamber. The dissolved air in excess of saturation at atmospheric pressure is released in fine bubbles that attach to oil and suspended particles and float them to the surface. There the solids can be skimmed off.

The other DAF type, vacuum flotation, is the application of partial vacuum to wastewater that has been saturated with air. The vacuum decreases the solubility of the air, which is released as fine bubbles. Vacuum flotation is more difficult to maintain and generally produces an effluent of poorer quality than the former method. Thus, it is used less frequently.

Dissolved air flotation can further be delineated in industrial waste treatment by low- and high-rate units. Low-rate DAFs have been used most often for separating light solids in industrial wastewaters. These systems typically are operated at surface loading rates of 0.7 to 1.4 $L/m^2 \cdot s$ (1.0 to 2.0 gpm/sq ft).

High-rate DAF systems are used extensively in the food industry and other industries where rapid recovery of floated solids is desirable. In the food industry, floated solids, oil, and grease are typically recovered for rendering into other byproducts. However, these solids decompose rapidly, requiring rapid collection and recovery. High-rate systems have smaller flotation basins than low-rate systems, allowing floated solids to be recovered quicker. Also, most high-rate systems have an additional thickening device as part of the solids removal mechanism, producing higher float solids (8 to 12%) than low-rate systems, which typically do not provide additional thickening. High-rate systems also are shallower than low-rate systems to reduce the rise time necessary to capture the sludge. Typical depths are 1.5 to 1.8 m (5 to 6 ft). High-rate systems normally require equalization ahead of the process to smooth out large fluctuations in flow that are common to food and other industries.

Chemical addition typically precedes sedimentation and flotation processes and may enhance solids removal. Inorganic chemicals like alum and ferric chloride aid the removal of fine solids by precipitating phosphates and hydroxides and enmeshing the fine solids in the precipitated solids. Some organic chemicals promote flotation of the suspended solids by altering the surface properties of the solids, the liquid, or the air bubbles, thereby improving adhesion of the air bubbles to the suspended solids.

If chemical addition is used to aid in floc formation, air injection via a pressurized effluent recycle stream may be desirable. After compressed air is added to the influent, the flow is held in the retention tank long enough to dissolve the air. The wastewater then flows through a pressure-reducing valve to a flotation tank equipped to collect and remove coarse solids that settle out, as well as those that float.

The addition of chemicals, while improving solids and fats, oils, and grease removal, generates substantial quantities of sludge that can be costly and difficult to process and dispose. In certain cases, where pretreatment standards are not stringent or where solid–liquid separation is easily achieved, air-only DAF systems are used to meet the desired results.

FILTRATION. Filtration may be used in pretreatment for suspended solids removal, especially when a low concentration of effluent suspended solids is desirable. Filtration of primary effluent removes suspended solids in quantities approximately equivalent to the noncolloidal fraction of the applied suspended solids. Filtration of industrial wastewaters is more often practiced downstream of other pretreatment processes than as a stand-alone pretreatment method.

Two processes with which filtration is commonly used are

- Neutralization/precipitation of heavy metals and
- Biological treatment to decrease BOD levels.

In both cases, filtration is used to polish suspended solids from the effluent to low levels before discharge to the WWTP.

Filtration is often used by the metal-finishing industry and printed-circuit-board manufacturers to further decrease heavy metal concentrations in effluent by capturing those metal hydroxide or sulfide solids that escape the sedimentation process. Filtration has also been used as an add-on process to neutralization/precipitation systems when metal pretreatment standards have been made more stringent.

Filtration of biological system effluent is most common for those industries with direct discharges to receiving streams, rather than for pretreaters. Polishing of biological effluent reduces both suspended solids and insoluble BOD.

Both granular media filtration and precoat filtration are discussed below. A more detailed discussion of filtration is available in *Design of Municipal Wastewater Treatment Plants* (WEF, 1992).

Granular Media. Granular media filters can incorporate single-size or graded media, continuous or interrupted operation, and a variety of flow patterns. The use of dual and multimedia configurations has been the predominant wastewater filtration practice in the U.S. These systems consist of two or more media—such as anthracite, sand, and garnet—with different specific gravities. The filters may be designed with different media intermixing zones that have a gradation of materials with different specific gravities. The theoretical advantage of a multimedia unit is the uniform decrease in pore space with increasing filter depth. The size and depth of the anthracite, sand, and garnet changes as a function of loading and strength of solids to be removed. Allowable hydraulic loading varies inversely with the anticipated suspended solids loading. High backwash rates are required for these types of filters.

Design considerations include hydraulic loading rates, media type, backwash practice, collection and distribution system, controls, and head loss development. Pilot studies are recommended to determine the filterability of the wastewater. The maximum hydraulic filtration rate is based on the sum of the peak plant flow and the design hydraulic and backwash flow from a filtration system. Hydraulic loadings average 1.8 to 6.8 $L/m^2 \cdot s$ (2.7 to 10 gpm/sq ft). Filter cleaning rates vary between 10 and 14 $L/m^2 \cdot s$ (15 and 21 gpm/sq ft).

Media size determines the penetration of solids into the filter. If the media are too large, the filtrate is of poor quality; if too small, solids accumulate on the surface, shortening the filter run. Round-shaped media are preferred, because they tend to rotate during backwash, scouring adjacent grains and freeing adhered solids more easily. Filter depth should be 50 to 100% greater than penetration depth. The finest media should have a depth of 150 mm (6 in.) or greater and a particle size of 0.35 mm (0.014 in.). Coarse media should be no larger than 2.0 mm (0.08 in.).

For cleaning and backwashing, typically up to 10% expansion of the finer particles should be provided. Conventional U.S. filtration practice requires total backwash flow of 3 100 to 4 100 L/m^2 (75 to 100 gal/sq ft), a loading that is independent of the rate of backwash. In addition, air scouring is frequently recommended, as is surface and subsurface washing or agitating equipment for mud ball removal. Backwash water generally is filter effluent.

The backwash water contains the solids and coagulant removed by the filter and, therefore, must receive further treatment before discharge. If clarifiers are present upstream of the filter, the backwash water may be routed through the clarifiers after equalization. Equalization may be provided for the raw wastewater stream or by separate tankage for the filter backwash alone.

If equalization and clarification are not components of the treatment process upstream of the filters, a decant tank with a bottom sloped to a sludge drawoff point should be provided. The decant tank would be operated in a semibatch mode with enough volume to allow time for settling of the backwash solids (30 minutes to 2 hours). After settling, a constant flow rate is pumped back to the rapid mix tank in the coagulation step before the filter. The size of the decant tank typically is based on receiving a full backwash volume with additional volume equal to the space occupied by settled sludge from previous and current backwash, plus a 0.6-m (2-ft) buffer zone above the settled sludge. Therefore, the size and configuration of the decant tank includes consideration of the frequency of sludge removal. If this is done manually once per day, the settled sludge volume expected from peak day influent loadings (suspended solids and coagulant) should be accommodated. Sludge removed from the decant tank is processed through thickening, storage, and dewatering either separately or in conjunction with other sludge produced from the pretreatment and source control systems.

One filtration technology that has wide industrial application uses a single-medium unstratified bed and continuous backwash (see Figure 5.10). This type of filter does not require specific operator attention for filter and backwash cycles, nor does it require a backwash equalization tank because backwash is continuous and at a constant rate.

Filter controls may be local manual, remote manual, or fully automatic. The control signal, either head loss or effluent turbidity, may be used to take a filter out of service, wash it, filter to waste until effluent quality is satisfactory, and return the filter to service. Conventional systems incorporate an underdrain system for collection of the filtrate and distribution of the wash water. Additionally, the unit typically incorporates a small supply of air introduced to the airlift chamber located in the center of the bottom of the tank. The design also should allow for necessary coagulant addition.

Precoat Filtration. A precoat filter typically is a rotary drum filter modified for filtration of wastewater containing fine or gelatinous solids that would otherwise plug the filter. The filter media are coated with a layer of porous filter aid such as diatomaceous earth. The wastewater is then filtered, leaving a thin layer of solids. The solids and a thin layer of filter aid are scraped off the drum, continually exposing a fresh surface of porous material to the wastewater.

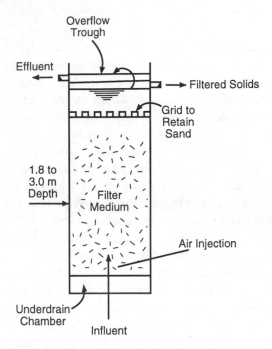

Figure 5.10   Continuous backwash upflow filter.

A precoat filter may operate under pressure. With the pressure type, discharged solids and filter aid collect in the housing and are removed periodically at atmospheric pressure while the drum is being recoated with filter aid.

SLUDGE HANDLING AND PROCESSING

Once sludge has been generated in a pretreatment process, it is often necessary to further process this sludge before disposal. The required steps may include thickening, stabilization, conditioning, dewatering, heat drying, and reduction. In most cases, not all of these processing steps are needed. The steps required depend on the characteristics of the individual sludges and the planned disposal method. Sludge handling warrants careful consideration because the costs associated with sludge processing and disposal can constitute 50% or more of the total treatment cost. Sludge handling options available during pretreatment processes are illustrated in Figure 5.11. Residual materials entering this handling scheme are secondary (or biological) sludge, chemical sludge, or other materials removed at the head end of treatment systems (for example, grit, scum, and screenings).

Handling options for pretreatment sludge must distinguish between granular-type sludge that thickens and dewaters easily to solids contents of 30 to 50% or greater and sludge with high water retention characteristics that

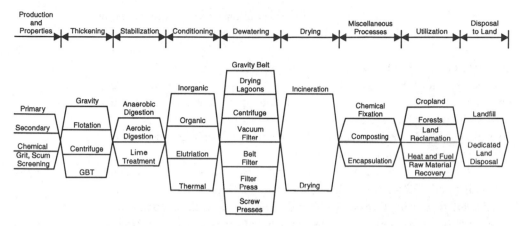

Figure 5.11 **Classification of sludge-handling and -disposal options commonly available for pretreatment systems.**

are difficult to dewater to a solids content greater than 20%. Biological sludge and many alum sludges are typical of the latter category and considerably more difficult to thicken and dewater than granular sludge.

Selection of sludge-handling alternatives should involve the consideration of

- Sludge properties, including water retention characteristics, oil and grease content, and hazardous or nonhazardous nature;
- Quantity of sludge;
- Solids content of sludge entering the sludge handling sequence;
- Local costs of land, energy, and labor;
- Local availability of land and land-based disposal options; and
- Local, state, and national regulations governing disposal.

The selection of appropriate sludge-handling processes begins with the identification of a cost-effective, locally available disposal option. It is not uncommon for sludge-handling processes to be selected, designed, and partially constructed before a disposal site has been selected. In such cases, sludge-handling processes will rarely be compatible with the disposal alternative in terms of the dewatered solids content of the sludge. Consequently, either the sludge is not dewatered sufficiently, or the sludge cake is drier than necessary for disposal. In addition, variables such as leachability of toxic materials, the degree of stabilization, and the presence of other objectionable constituents may render the sludge unacceptable for available disposal alternatives.

In the preliminary selection and design of sludge-handling alternatives, it usually is wise to conduct treatment tests to identify conditioning requirements and the performance that can be expected from each unit process. Even in cases where few problems are anticipated with the performance of

---

the mainstream treatment processes, minor variations in waste stream composition can significantly alter the character of the sludge produced.

SLUDGE CONDITIONING. Most sludge requires some sort of conditioning before thickening and dewatering. Conditioners may be divided into two categories: inorganic coagulants and synthetic organic polyelectrolytes. In addition, inert materials such as fly ash, cement kiln dust, sawdust, carbon-based byproducts, or other industrially generated materials may be used to absorb water and increase the structural stability of the sludge.

Inorganic coagulants like lime, alum, and ferric salts are effective in coagulating colloidal sludge particles and forming an inorganic matrix in organic sludge, making it easier to dewater. However, coagulants contribute substantially to the volume of sludge, for which handling and disposal is necessary. Synthetic polyelectrolytes, which are generally used in lower concentrations than inorganic coagulants, may be effective in increasing the particle size of the sludge, thus aiding in thickening and dewatering. In many cases, inorganic coagulants and polyelectrolytes are used in combination to optimize dewatering while reducing the cost of conditioning chemicals. If land disposal or incineration is practiced, the use of ferric chloride or lime should be carefully evaluated to avoid problems involving salinity and corrosion. Bench-scale laboratory tests are almost always necessary to determine the optimum dose and type of conditioning agent or agents required and the sequence in which these must be added to condition the sludge.

SLUDGE THICKENING AND DEWATERING. Two methods for separating solids in streams containing suspended solids concentrations greater than approximately 1% are thickening and dewatering. Typical unit operations for each are

- Thickening—gravity thickening, dissolved air flotation (DAF) thickening, centrifugal thickening, and gravity belt thickening; and
- Dewatering—pressure filtration, belt filtration, vacuum filtration, centrifugation, screw presses, sand bed drying, and drying lagoons.

Thickening generally refers to those processes resulting in solids streams that behave as liquids, whereas dewatering processes usually produce streams too thick to behave as liquids. Generally, dewatering performance can be improved when streams with solids concentrations less than 5% are thickened before dewatering. Dewatering operations should yield solids concentrations of at least 15% and as high as 50%.

Thickening. Thickening processes are used to reduce sludge volume before dewatering or disposal. In most cases, the performance of a sludge-dewatering process improves substantially as the feed solids concentration to that process increases. Therefore, improving the performance of the thickening process usually reduces the cost of subsequent dewatering. For instance, calcium carbonate sludge can rarely be effectively dewatered at feed solids concentrations less than 6 to 7%. Increasing the feed solids content above this concentration, however, greatly improves the dewatering rate. In one case, a feed concentration of approximately 10% resulted in a solids filtration rate of 50 to 60 kg/m^2·h (10 to 12 lb/hr/sq ft), whereas thickening the sludge to 20 to 22% solids content increased the filtration rate to 240 to 290 kg/m^2·h (50 to 60 lb/hr/sq ft) (Okey *et al.*, 1979).

GRAVITY THICKENING. Gravity thickening of sludge is accomplished in a clarifier-type settling tank with a slowly rotating rake mechanism (Figure 5.12). A picket rake mechanism is used to break up sludge bridging and promote settling and compaction of the solids. The most important criterion in thickener design is the surface area required to effect the desired thickening. This area is calculated by establishing a limited solids loading rate or solids flux, expressed in units of kilograms dry solids fed per day per square metre of thickener surface area (pounds of solids per square foot per day). The most commonly used procedure was developed by Dick (1970). A design procedure for this process was outlined in *Process Design Techniques for Industrial Waste Treatment* (1974).

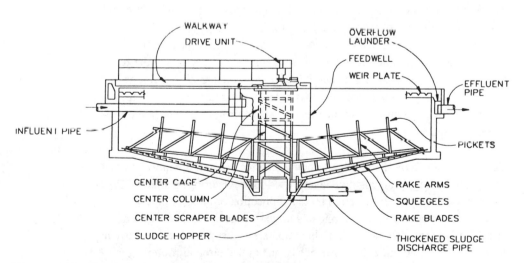

Figure 5.12   Typical gravity thickener (WEF, 1992).

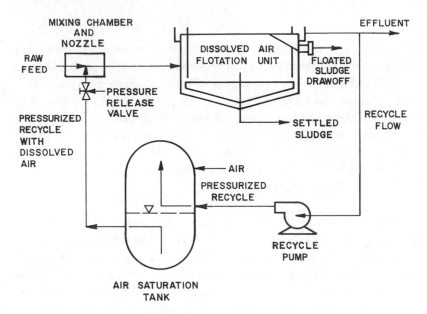

Figure 5.13   Schematic of operating dissolved air flotation system with recycle flow pressurization.

DISSOLVED AIR FLOTATION. Dissolved air flotation thickening of solids, as shown in Figure 5.13, is achieved by pressurizing liquid (in sludge-thickening applications, this stream is usually the clarified subnatant) and mixing the resulting product with the influent sludge stream. When the combined stream is released to atmospheric pressure, small air bubbles form and become attached to and enmeshed in the sludge flocs. The attachment of gas bubbles to the solids reduces the specific gravity of the combined mixture, allowing the solids to float to the surface where they are skimmed from the clarified wastewater in a concentrated form. The pressurized sludge stream is released to a circular or rectangular flotation tank.

Design variables include the air-to-solids ratio required for effective thickening of the sludge, solids loading rate, hydraulic loading rate, and the feed solids concentration. Generally, DAF thickening produces higher sludge concentrations than gravity thickening, particularly on lighter or thinner sludges. However, DAF is a more complex process mechanically and has greater operational requirements.

CENTRIFUGATION. In addition to being an effective unit operation for sludge thickening, centrifugation also is used for dewatering of sludge. The process accelerates sedimentation by the application of circular motion. Three types of centrifuge are applicable to wastewater treatment: solid bowl scroll, basket, and disk nozzle. All three operate on the principle of removing solids from a liquid under the influence of an applied circular motion, generally in the range of 1 000 to 6 000 times the force of gravity. The fundamental difference in the three types is the method by which solids are collected and discharged from the bowl. A schematic illustration of a solid bowl centrifuge is shown in Figure 5.14. Additional details regarding the types of centrifuges and design procedures for each are available in WEF (1992), U.S. EPA (1979a), and *Process Design Techniques for Industrial Waste Treatment* (1974).

The principal advantage of centrifugal thickening is the relatively high solids content that can be achieved. Also, centrifuges can handle large sludge flows, requiring fewer units in large applications. Generally, this is not an issue in industrial pretreatment applications. Typically, a solids content significantly in excess of that achievable using gravity thickening and somewhat in excess of that attainable using DAF thickening can be achieved, depending on the sludge characteristics. The principal disadvantages of centrifugation include high energy costs and required maintenance because of abrasion, especially in solid bowl centrifuges. Plugging also can be a problem with some sludge types in disk-type centrifuges.

In industrial applications, centrifuges have been particularly favored for dewatering applications in processing paint sludge from automotive and furniture-finishing operations. Also, so-called "3-phase" centrifuges are used for processing oily wastewater solids into recoverable oil, dewatered solids, and water. This special application is discussed in more detail later in this chapter.

GRAVITY BELT THICKENING. The gravity belt thickener is a popular and cost-effective thickening device. Essentially, this technology consists of the gravity dewatering section of a belt filter press used in sludge dewatering. Conventional coagulation and flocculation techniques are used to free bound water from the sludge. The bound water drains through the permeable belt to

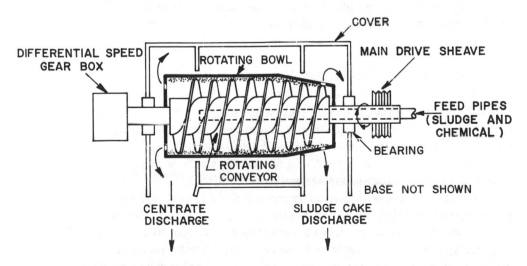

Figure 5.14 Continuous countercurrent solid bowl conveyor discharge centrifuge.

produce thickened sludge concentrations of 4 to 8%. Gravity belt thickeners generally are less costly and complex than many of the competing technologies and produce comparable thickening results. The units are typically sized on a solids loading basis, with design loading determined by the nature and solids content of the sludge to be thickened.

Dewatering. Reducing the moisture content in the sludge is essential regardless of whether disposal is by incineration or landfill. Typically, sludge is thickened before application to dewatering processes. Several more common dewatering methods include centrifugation (discussed in the previous section), vacuum filtration, belt filtration, pressure filtration, screw presses, and drying beds. Although many plants designed in the past use vacuum filtration, this equipment has largely been replaced by processes that typically achieve better cake solids content, have lower electrical operating costs, and are mechanically less complex to operate and maintain.

BELT FILTER PRESSES. Several belt filter presses were introduced to the U.S. market during the 1970s. Most belt filters consist of three stages: a chemical mixing and conditioning system, a section in which the sludge drains by gravity to a nonfluid consistency (similar to the gravity belt thickener), and a section in which compaction further dewaters the sludge (Figure 5.15). The chemically conditioned sludge is applied to a moving belt. In this state, gravity drains water from the sludge. Next, the sludge rolls beneath a second belt to be further compacted. In this section, the tension between the belts is increased and shearing action is applied to the sludge by passing it along a tortuous path between rollers. As the sludge leaves the press, the two belts separate and allow the sludge to discharge to a hopper. Important parameters in the design of belt filter presses include chemical conditioning requirements, hydraulic and solids loading limitations, and belt washwater requirements.

RECESSED-PLATE FILTER PRESSES. Sludge filtration using recessed-plate filter presses is applicable when a relatively high solids content in the dewatered sludge is required. This is a popular technology in industries that use heavy metals (for example, plating and printed-circuit board manufacturing), and pay a premium to dispose of sludge in a secure landfill.

Pressure filtration is a batch dewatering operation in which the sludge is applied to the filter by high-pressure pumps. Water is driven from the sludge through the filter media in response to the pressure applied by the feed pump. Filtrate flows from behind the filter media through flow passages formed on the plate surface or in open mesh underdrainage chambers to outlets in the filter plate. After the cake is completely formed at the final filtration pressure, the cycle is ended and the filter chamber is opened, one chamber at a time, allowing discharge of the filter cake to a hopper or conveyor belt. For sludge

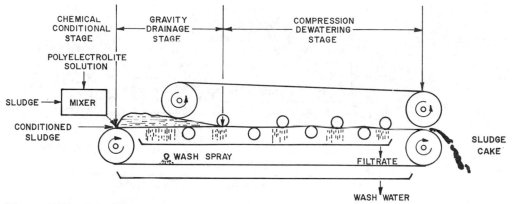

Figure 5.15   Belt filter press.

that is difficult to dewater, a diatomaceous earth precoat may be applied to the filter media to increase the achievable cake solids content. These systems operate at pressures ranging from 340 to 1 550 kPa (50 to 225 psig).

The primary advantage of pressure filtration is the high cake solids content achievable. Ordinarily, pressure filtration achieves the driest cakes obtainable using conventional sludge dewatering processes, excluding those that also accomplish some drying of the sludge. Disadvantages include the high cost of the system, high requirements for operator time because of the batch nature of the process, and resulting high dewatering costs.

SCREW PRESSES. Screw presses typically are used to dewater sludge produced by treating pulp- and paper-manufacturing process water for reuse. Where high volumes of biological solids are present in the feed, screw presses exhibit diminished performance.

The raw sludge feed typically is thickened using polymers, or in some cases alum, before the screw press process. Thickened feed solids concentration of 7% is common, while unthickened feeds range from approximately 1 to 3% (Busbin *et al.*, 1991; Feaster and Keenan, 1990; and NCASI, 1991). Feed sludge enters the screw press via the inlet feed (see Figure 5.16). Sludge moves from the low pressure (or gravity drainage) zone to the high-pressure zone by the action of the screw rotating at 0.5 to 6 rpm. The change in pressure gradient is controlled by the backpressure generated against the screw, which is controlled by the choke plate, located at the discharge end of the screw. Some screw presses have a steam injection port at the inlet end of the unit. The steam is pressurized to 7 to 345 kPa (1 to 50 psi) and supplied to a heat jacket (the screw itself) to improve dewaterability of the feed sludge (NCASI, 1991).

Cake solids of 30 to 60% have been attained using this technology, with an average of 42% being typical. Solids capture rates vary, with averages of

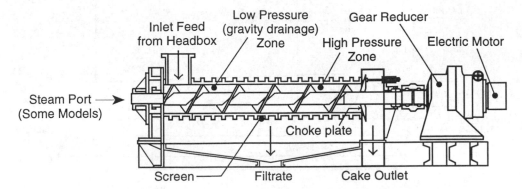

Figure 5.16   Schematic diagram of a typical screw press.

93 to 95% being common (Busbin *et al.*, 1991, and NCASI, 1991). Polymer dosages vary widely, with systems reporting values of 2 to 10 kg/ton of solids applied (4 to 22 lb/ton) (Busbin *et al.*, 1991; Feaster and Keenan, 1990; and NCASI, 1991). Alum dosages also vary greatly, from 3 to 36 kg/ton of solids applied (7 to 79 lb/ton) (Busbin *et al.*, 1991).

The main advantage of the screw press process is the high solids content of the resultant cake. This results in lower transportation and disposal costs. The disadvantages are high initial capital costs and greater operational control required by plant staff.

SAND DRYING BEDS. For industries in which a relatively small amount of sludge is generated, sand drying beds may be an attractive alternative for dewatering where capital cost is the major concern. The principal difference between sand drying beds and mechanical forms of sludge dewatering is related to the significant influence of climatic conditions and design and performance of the system. The sludge is dried by percolation of water through the sludge and supporting sand bed and by evaporation of water from the sludge surface. The proportion of water removed by percolation may vary from 20 to 55%, depending on the initial solids content of the sludge and the water retention characteristics of the solids.

Wet sludge is typically applied to the beds at a depth of 200 to 300 mm (8 to 12 in.). The sludge is allowed to dewater and dry on the bed until it can be removed manually by pitch forks or mechanically by front end loaders. Chemically conditioning the sludge before application to the beds often significantly reduces the time required for the sludge to dry. A rational method for the design of sand beds based on observed drainage characteristics of the sludge and climatic conditions has been developed and summarized as a design procedure (Process Design, 1974).

SLUDGE DRYING. Reducing sludge volume has become increasingly important in certain industries, as both hazardous and industrial sludge disposal costs have increased. Another important consideration that has led to the use of the sludge drying process for biological sludge is that drying has been designated by the U.S. Environmental Protection Agency (U.S. EPA) as a process to further reduce pathogens (PFRP).

To meet this designation, the dewatered sludge cake must be heated to higher than 80°C (176°F), or the wet bulb temperature of the gas stream in contact with the sludge at the point where it leaves the dryer must be higher than 80°C. Further, sludge moisture content must be reduced to 10% or less. Industrial biosolids receiving the PFRP designation and meeting metals and nutrient criteria for land application typically can be applied under most beneficial use programs. Three industries that use sludge drying are the pulp and paper industry, the metal-finishing industry, and the specialty steel industry.

In the case of the pulp and paper industry, the high volatile content of the wastewater solids allows dried wastewater solids to be burned with bark or other solid fuel for steam and energy. Dried biosolids are also land applied to forest land as part of beneficial use programs.

The metal hydroxide sludge produced by metal-finishing and steel industries are often hazardous and, therefore, have high disposal costs. Reduction in overall sludge volume by drying reduces both handling and disposal costs.

There are two major methods used for drying sludge:

- Direct-fired convection and
- Indirect heat.

With the direct-fired method, heated air or flue gas is passed over the sludge to evaporate water. The resulting gas–vapor mixture is either discharged directly to the atmosphere or, more commonly, scrubbed or condensed and returned to the wastewater treatment system.

Indirect heat drying refers to the use of a heat exchanger and a heat source such as steam, thermal oil, or hot air. The heat source does not come into direct contact with the sludge in indirect heat drying.

Both types offer advantages and disadvantages depending on the nature of the sludge (solids content, volatile materials present) and the availability and cost of energy. Direct-fired dryers are more efficient but create more offgas and vapor than indirect dryers. When hazardous volatile materials are present, offgas treatment costs may dictate the use of indirect dryers. Also, indirect systems recover heat from the process, reducing overall energy input.

The principal types of direct-fired dryers are

- Rotary-driven,
- Flash,
- Tray,

- Fluid-bed, and
- Belt.

The principal types of indirect dryers are

- Thin-film,
- Screw,
- Paddle, and
- Disc.

In a thin-film drying system, dewatered sludge is pumped to the first stage of a two-stage dryer. With a typical residence time of 5 to 10 minutes, dewatered sludge is dried to 40 to 65% dry solids. The second stage further decreases the moisture content of the sludge. With a 1- to 2-hour residence time, the second-stage dryer produces a pelletized or granular product that is more than 90% dry solids. Water evaporated from the sludge is condensed and returned to the wastewater treatment process.

An example of the volume and weight reduction possible from sludge drying is shown in Table 5.1.

A more thorough discussion of thermal drying technologies is found in WEF (1992).

SLUDGE COMPOSTING. Composting is a biological stabilization process used with certain in-process industrial wastewater solids from sidestreams for which beneficial use is possible. Typically, composting has been most applicable to various sludges generated in the food, pharmaceutical, and pulp and paper industries.

In the composting process, dewatered sludge is mixed with a bulking agent such as wood chips, sawdust, or previously composted biosolids and allowed to further decompose. Composting is an aerobic biological process requiring a continuous supply of oxygen, provided either by frequent turning or

Table 5.1 Weight/volume reduction from dewatering and drying a typical biological sludge.

	Percent solids	Gallons	Wet tons	Dry tons	Bulk density, lb/cu ft[a]	cu ft[b]
Feed sludge	2	10 000	42.0	0.8	62.4	1 337
Mechanical dewatering	23	870	3.6	0.8	64	113
Dryer	90	222	0.9	0.8	50	37

[a] lb/cu ft $\times$ 16.02 = kg/m^3.
[b] cu ft $\times$ (2.832 $\times$ 10^{-2}) = m^3.

mechanical blowers. The bulking agent facilitates the uniform passage of air through the pile.

The microbial decomposition generates its own heat, creating temperatures of 55 to 60°C (131 to 140°F). The original weight of the sludge typically is reduced in the composting process by more than 50% through evaporation and destruction of biodegradable solids.

The three most common types of composting systems are

- Windrows,
- Aerated static piles, and
- Enclosed mechanical systems.

Enclosed mechanical systems are increasingly popular for municipal wastewater solids but relatively expensive and complex, and not in wide use in industry.

In windrow composting, mounds of the wastewater solids-bulking agent mixture are turned approximately every 2 days by specialized equipment to aerate the pile and expose the entire pile to microbial action. No mechanical aeration is used.

Aerated static pile composting uses aeration piping placed under the wastewater solids-building agent mixture. The pile is not moved for the entire 14- to 21-day aeration time. Oxygen is provided to the pile by mechanical blowers that either draw air through or discharge air into the pile.

Achieving a proper nutrient balance is a major concern in composting industrial biosolids. Generally, a carbon-to-nitrogen (C:N) ratio of 20 to 30:1 is considered optimum. In municipal biosolids, this high C:N ratio is achieved by adding the bulking material. In industrial biosolids, supplemental nitrogen may be necessary to achieve the optimal C:N range. Also, toxic organic compounds can inhibit the composting process and require evaluation by pilot testing before full-scale composting.

A detailed evaluation of composting pulp and paper biosolids using an aerated static pile is provided by Campbell *et al.* (1991). Two successful composting studies for the food industry (one an enclosed-mechanical system on poultry-processing waste, the second a windrow composting system for seafood waste) are discussed in Carr *et al.* (1990). Detailed design considerations for each type of composting system, including land, bulking agent, and nutrient requirements are described for municipal biosolids in WEF (1992).

DISPOSAL PRACTICES AND TECHNOLOGY

GRIT AND SCREENINGS. Items typically collected on coarse screens include rags, pieces of string, lumber, rocks, leaves, plastics, and other materials representative of the manufacturing operations in a particular industrial plant. Grit generally consists of heavy, coarse solids like mill scale, gravel, sand, cinders, nails, and bottle caps. In addition, large particles of organic matter frequently may be removed in grit. This is particularly true in the fruit and vegetable industry, where water is used to both wash and move the product from delivery to processing. Field dirt, vines, leaves, and damaged product are often found in the grit and screenings from this industry.

Although washing grit can reduce the amount of organic matter present, it is rarely possible to completely separate inorganic and nonputrescible materials from organics that cause odor problems. Depending on the type of industry and waste, grit and screenings may be odorous and tend to attract rodents and insects. In addition, grit can be extremely abrasive and hard to handle in mechanical treatment systems.

The most common disposal methods for screenings and grit are landfilling or incineration. Incineration of grit and screenings keeps these materials out of other sludge, reduces the volume requiring disposal, and results in the destruction of any pathogens that may be present in these residuals. Depending on the industry, screenings and grit may contain some putrescible materials and, if landfilled, should be covered at sufficient frequency to meet requirements. Sprinkling with lime may reduce odors from temporarily uncovered solids. Because residues from the incineration of grit and screenings may contain relatively high concentrations of trace metals, special precautions are required in some locations to landfill the incinerator ash.

A more detailed consideration of the handling and disposal of grit and screenings is contained in U.S. EPA's *Process Design Manual for Sludge Treatment and Disposal* (1974) and in standard design texts (WEF, 1992, and Conway and Ross, 1980).

OILY SLUDGE AND RESIDUES. The handling and disposal of oily sludge has been discussed in Adams and Koon (1977), Adams and Stein (1976), API (1969), and Scher (1976). Conditioning, thickening, and dewatering of oil-laden sludge vary significantly, depending on the source. Sometimes, sludge-handling processes may be capable of separating sludge from suspended solids or chemical coagulants mixed with the sludge. However, in most cases, the dewatered wastewater solids contain a considerable quantity of oil. Disposal methods include recovery, incineration, landfilling, and landfarming.

Oily sludge is a particular problem in the steel industry, refineries, petrochemical plants, and food-processing plants. The presence of excess amounts of oil often prevents the wastewater solids from being land-applied in a beneficial use program or buried in a conventional landfill.

Oils recovered for reuse must be separated from any suspended solids when the solids constitute more than 1% of the wastewater solids–oil mixture. Separation may be achieved through conventional clarification by heating the oil at approximately 88°C (190°F) for 4 to 6 hours, followed by settling for 12 to 24 hours. At the end of this period, the clean oil will have risen to the surface. A middle layer of secondary oil emulsions and a bottom layer of water-containing soluble oil components, suspended solids, and oils also exist. These layers must be reprocessed or disposed of by other methods. The American Petroleum Institute's *Manual of Disposal of Refinery Wastes* (1969) contains additional details regarding recovery techniques for oily refinery wastewater solids. Information on requirements for sludge containing hazardous constituents is also available in API (1984).

One effective method of separating and recovering oil from wastewater solids is the 3-phase centrifuge. A typical 3-phase centrifuge system is shown in Figure 5.17. In this system, oily wastewater solids are pumped to a heat tank where steam is injected directly, raising the wastewater solids temperature to 82 to 93°C (180 to 200°F). The wastewater solids are fed to a special decanter-type centrifuge, then processed on a single pass into three separate phases: oil, water, and dewatered solids. Typically, the oil is recovered for fuel blending or further refining, or in the case of food waste, is sold to the rendering industry. The solids may be sufficiently free of oil to permit lower-cost disposal options, such as conventional landfills or land application. Again, the food industry often finds that high-protein solids can be sold or given to renderers.

Incineration of oily sludge and subsequent landfilling of the ash may provide an acceptable means of disposal. Incineration may be accomplished using fluidized bed, rotary kiln, and multiple-hearth furnaces. In addition, waste

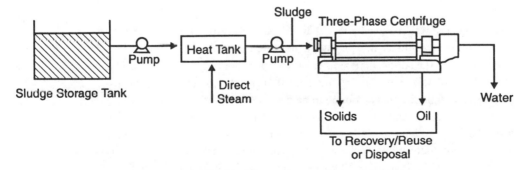

Figure 5.17 Schematic of a typical three-phase centrifuge system for handling oily sludge.

oils that are relatively free of suspended material can be incinerated in liquid burners, provided waste disposal regulations are met. The fluidized bed incinerator is best suited for sludge that is partially liquid because the incinerator may be fed by pumps and screw conveyors. The multiple-hearth incinerator is preferable if the feed is in the form of a cake or nonpumpable solids. Landfilled waste oils and oily sludge must comply with regulations promulgated under the Resource Conservation and Recovery Act (RCRA). The disposal of oily wastewater solids on soil is acceptable only if such disposal does not contaminate groundwater or stormwater runoff or create a potential seepage problem. A proper landfarming operation using soil bacteria for degradation of oils can satisfy these requirements. Landfarming of oily biosolids has been successfully practiced by refineries where sufficient land area is available for proper decomposition of the oil-containing solids (Kincannon, 1976, and Huddleston and Cresswell, 1976).

Landfarming involves spreading the biosolids in 100- to 150-mm (4- to 6-in.) layers, allowing the biosolids to dry for approximately 1 week, adding nutrients, and then disking the biosolids into the soil. Decomposition rates have been found to average 8 kg/m^3/month (0.5 lb/cu ft/month) without nutrient addition and 16 kg/m^3/month (1.0 lb/cu ft/month) with nutrient addition.

SLUDGE CLASSIFIED AS TOXIC OR HAZARDOUS WASTE. Sludge from the treatment of industrial wastewater may be classified as hazardous or toxic according to national or local environmental regulations. Some wastewater treatment sludge is classified as listed hazardous wastes by U.S. regulations promulgated under RCRA. Others may be classified as characteristic hazardous wastes because they possess the characteristics of ignitability, corrosivity, reactivity, or toxicity (as determined by the toxicity characteristic leaching procedure). Neither characteristic nor listed hazardous wastes may be land-disposed without prior treatment to specific standards (see 40 CFR Part 503 sludge regulations). A detailed discussion of technical and legal requirements related to hazardous waste disposal is beyond the scope of this book; the reader is referred to Lindgren (1989).

NONHAZARDOUS WASTEWATER SOLIDS. Landfilling. Landfilling is the planned burying of refuse, sludge, or other materials. This process involves placing the solids in a prepared site or excavated trench and covering with a layer of soil. In evaluating the landfill disposal of material, it is necessary to determine the waste volume to be disposed of; production rate; and physical, chemical, and engineering properties of the waste material. This information is necessary for calculating spatial needs, estimating cover requirements (if any), determining personnel and equipment needs, and determining the landfill design. In addition, the candidate waste for land disposal must have sufficient structural stability to be worked conveniently with conventional earth-moving equipment. Operations that increase the structural stability of waste materials include dewatering, drying, mixing with dry, absorbent materials, chemical fixation, or a combination of these methods.

There are three basic types of landfill:

- Trench landfill—Trenches are used where groundwater is well below the surface so that a large trench or hole may be dug without intersecting the water table. Soil that has been removed from the trench is typically used as cover.
- Area landfill—The material to be landfilled is spread and compacted on an existing ground surface with cover soil spread and compacted over the waste to achieve the completed cell design.
- Ramp landfill—The ramp method is a combination of the first two methods. Refuse is spread and compacted on a slope, then covered with soil that has been obtained by excavating the front of the working space. Usually, the area method requires cover material to be transported to the landfill site, whereas the other two methods use soil available at the site.

Detailed descriptions of various landfill methods can be found in other sources (Process Design, 1974; Tchobanoglous *et al.*, 1977; Pavoni *et al.*, 1973; and Duvel, 1979). It is important to comply with relevant permit or licensing requirements with respect to solid waste disposal. In the U.S., such permits typically are issued at the state level. In addition, performance standards typically cover such topics as construction standards and general site selection criteria.

Landfarming. Where adequate land is available within a reasonable distance from the industrial site, landfarming can be an attractive disposal method. Candidate wastes for landfarming should contain biodegradable constituents and should not be subject to significant leaching while the degradation proceeds. Food biosolids and pharmaceutical biosolids are good candidates for this type of beneficial reuse. Biosolids may be landfarmed in either liquid form (2 to 5% solids) or dewatered cake (18 to 25% solids).

In a land application system, biosolids are applied to the soil. Microorganisms present in the soil oxidize constituents present in the biosolids and, in turn, contribute to the soil's organic content. Nutrients present in the applied biosolids contribute to the growth of crops on the land. Biosolids typically are applied to the land by spraying, subsurface injection, or spreading in manure-type spreaders if the biosolids are sufficiently concentrated. The principles and design concepts of land application systems have been described in several texts (Overcash and Pal, 1979; Loehr *et al.*, 1979; and Huddleston, 1979).

Incineration. Incineration is a process that uses thermal oxidation to convert the waste to an inert ash product and gases. In doing so, the process significantly reduces bulk and toxicity, as well as the potential for decomposition. There are several types of incineration devices, each of which has distinct advantages and disadvantages. The processes and their principal uses have been described by Conway and Ross (1980), Hitchcock (1979), and WPCF (1988). Although incineration can be costly compared to landfilling or landfarming, the increasing cost of land disposal and concerns about the potential long-term liability associated with this make incineration attractive for disposal of sludge that is not of high metal content. Incineration of metal-containing sludge is likely to produce ash, for which disposal is difficult and costly because of its hazardous classification.

CHEMICAL WASTEWATER SOLIDS. Chemical wastewater solids generated most often from industrial wastewater pretreatment operations include those from neutralization systems in which significant quantities of precipitate are produced and those from chemical coagulation systems using inorganic salt coagulants. Methods that have been used for disposal of chemical wastewater solids include lagooning, landfilling, land application, and recovery of coagulants. The disposal of water treatment plant sludge (principally alum sludge) has received much attention (AWWA, 1978a; AWWA, 1978b; and Bishop, 1978).

Lagooning. Lagooning has historically been the most common method of chemical sludge disposal. Recently, concerns about long-term liability and tighter regulatory requirements have decreased the popularity of this disposal method. It is well documented that alum sludge is difficult to dewater by this method because of the high water-retention characteristics of the sludge. In most cases, alum sludge does not thicken to concentrations greater than 9% after several years of settling (AWWA, 1989, and AWWA, 1990). On the other hand, lime sludge dewaters much more readily in lagoons, with concentrations of 40 to 50% reported (AWWA, 1989, and AWWA, 1990). Odors from lagooning lime sludge are rare and, when they occur, are usually only temporary. Dewatering characteristics of lime sludge can be expected to vary greatly with the magnesium content of the sludge. Because magnesium hydroxide precipitates a fluffy floc, whereas calcium carbonate precipitates as a rather granular solid, achievable solids concentrations will probably be less when lagooning sludge containing high concentrations of magnesium. Calcium sulfate reacts similarly because it also forms granular precipitates. The supernatant of leachate from calcium sludge usually requires treatment because of its high pH.

Landfilling. Subject to regulatory requirements, nonhazardous chemical sludge may successfully be landfilled if the sludge has been sufficiently dewatered to remove all free water that would cause leachate problems and if it is sufficiently dry to have the required structural stability for the type of landfill to which it is applied.

Dewatered alum sludge has been disposed of by landfilling in Houston, Texas, and Atlanta, Georgia (AWWA, 1978a). In Houston, sludge has been deposited in a landfill dedicated to alum sludge disposal.

Land Application. Lime biosolids may be applied to land as a neutralizing agent in place of agricultural grade limestone. The higher level of nutrients in most lime biosolids adds fertilizer value not obtained with agricultural lime. Calcium sulfate biosolids may be applied on soils where gypsum is required for conditioning; however, high loadings are not used. Alum biosolids have been successfully applied indirectly to land at rates of 25 mm/a (1 in./yr), with no adverse effect on vegetation (U.S. EPA, 1979b). Elsewhere, clogging of the soil by aluminum hydroxide has reportedly caused a problem with vegetation growth (Sankey, 1967). After the soil has frozen and thawed, however, the biosolids are transformed from a gelatinous to a granular structure, thus alleviating clogging problems.

Recovery. Recovery of alum and lime wastewater solids often has been practiced, but not the recovery of calcium sulfate sludge and iron-based sludge (Sankey, 1967; Parker *et al.*, 1974). Both acidic and alkaline methods of alum recovery exist, the acidic recovery method being the more common. Typically, alum wastewater solids are incinerated to remove organics and convert the aluminum to aluminum oxide. Subsequently, ash is mixed with sulfuric acid to form aluminum sulfate. Culp *et al.* (1978) estimate that the cost of the acidic recovery method is approximately one-third the cost of buying fresh alum.

Lime can be recovered by calcining lime wastewater solids. This process involves incineration of the wastewater solids to convert carbonates and hydroxides to calcium oxide. Centrifugation has been used to partially separate recycled inert materials from the lime to improve the effectiveness of the process.

R*EFERENCES*

Adams, C.E., Jr., and Koon, J.H. (1977) *The Economics of Handling Refinery Sludges. Proc. Open Forum Manage. Pet. Refinery Wastewaters*, Univ. of Tulsa, Okla.

Adams, C.E., Jr., and Stein, R.M. (1976) Sludge Handling Methodology for Refinery Sludges. *Proc. Open Forum Manage. Pet. Refinery Wastewaters*, Univ. of Tulsa, Okla.

American Petroleum Institute (1969) *Manual of Disposal of Refinery Wastes.* Division of Refining, New York, N.Y.

American Petroleum Institute (1984) *The Land Treatability of Appendix VIII Constituents Present in Petroleum Industry Wastes.* Washington, D.C.

American Public Health Association. (1992) *Standard Methods for the Examination of Water and Wastewater.* 18th Ed., Washington, D.C.

American Water Works Association (1978a) Water Treatment Plant Sludges— An Update of the State of the Art: Part 1. *J. Am. Water Works Assoc.*, **70**, 9, 498.

American Water Works Association (1978b) Water Treatment Plant Sludges— An Update of the State of the Art: Part 2. *J. Am. Water Works Assoc.*, **70**, 10, 548.

American Water Works Association (1989) Sludge: Handling and Disposal. Publ. No. 20034, Denver, Colo.

American Water Works Association (1990) *Water Quality and Treatment.* 4th Ed., F.W. Pontius (Ed.), McGraw–Hill, Inc., New York, N.Y.

Besselievre, E., and Schwartz, M. (1976) *The Treatment of Industrial Wastes.* McGraw–Hill, Inc., New York, N.Y.

Bishop, S.L. (1978) Alternate Processes for Treatment of Water Plant Wastes. *J. Am. Water Works Assoc.*, **70**, 9, 503.

Busbin, S., *et al.* (1991) Sludge Dewatering at a 100 Percent ONP Furnish Mill. *Proc. Tech. Assoc. Pulp and Paper Ind., Environ. Conf.*, TAPPI Press, Atlanta, Ga.

Campbell, A., *et al.* (1991) Composting of a Combined RMP/CMP Pulp and Paper Sludge. *Proc. Tech. Assoc. Pulp and Paper Ind., Environ. Conf.*, TAPPI Press, Atlanta, Ga.

Carr, L.E., *et al.* (1990) Mechanical Composting of Dissolved Air Flotation Solids from Poultry Processing. *Proc. 1990 Food Ind. Environ. Conf.*, Ga. Tech. Res. Inst., 391.

Conway, R.A., and Ross, R.D. (1980) *Handbook of Industrial Waste Disposal.* Van Nostrand Reinhold, New York, N.Y.

Culp, R.L., *et al.* (1978) *Advanced Wastewater Treatment.* Van Nostrand Reinhold, New York, N.Y.

Dick, R.I. (1970) Thickening. In *Advances in Water Quality Improvement— Physical and Chemical Processes.* E.F. Gloyna and W.W. Eckenfelder, Jr. (Eds.), Univ. of Texas Press, Austin.

Duvel, W A., Jr. (1979) Solid Waste Disposal: Landfilling. *Chem. Eng.*, **86**, 14, 77.

Feaster, T.D., and Keenan, S. (1990) Chemical Optimization and Design of a Sludge Dewatering Facility for an Unbleached Kraft Mill. *Proc. Tech. Assoc. Pulp and Paper Ind., Environ. Conf.*, TAPPI Press, Atlanta, Ga.

Hitchcock, D.A. (1979) Solid Waste Disposal: Incineration. *Chem. Eng.*, **86**, 11, 185.

Hopwood, A.P. (1980) Recovery of Protein and Fat from Food Industry Waste Waters. *J. Inst. Water Pollut. Control*, 2, 225.

Huddleston, R.L. (1979) Solid Waste Disposal: Landfarming. *Chem. Eng.*, **86**, 5, 119.

Huddleston, R.L., and Cresswell, L.W. (1976) The Disposal of Oily Wastes by Land Farming. *Proc. Open Forum Manage. Pet. Refinery Wastewaters*, Univ. of Tulsa, Okla.

Kincannon, C.B. (1976) Oily Waste Disposal by Soil Cultivation. *Proc. Open Forum Manage. Pet. Refinery Wastewaters*, Univ. of Tulsa, Okla.

Lindgren, G.F. (1989) *Managing Industrial Hazardous Waste.* Lewis Publishing, Inc., Chelsea, Mich.

Loehr, R.C., *et al.* (1979) *Land Application of Wastes.* Vol. 1 and 2, Van Nostrand Reinhold, New York, N.Y.

Industrial Pollution Control Handbook (1971). H.F. Lund (Ed.), McGraw-Hill, Inc., New York, N.Y.

Metcalf & Eddy, Inc. (1991) *Wastewater Engineering: Treatment, Disposal, Reuse.* McGraw-Hill, Inc., New York, N.Y.

National Council of the Paper Industry for Air and Stream Improvement (1991) *Full-Scale Experience with the Use of Screw Presses for Sludge Dewatering in the Paper Industry.* NCASI Tech. Bull. No. 612, New York, N.Y.

Nemerow, N.L. (1978) *Industrial Water Pollution Origins, Characteristics, and Treatment.* Addison–Wesley Publishing Co., Reading, Mass.

Nemerow, N.L. (1991) *Industrial and Hazardous Waste Treatment.* Van Nostrand Reinhold, New York, N.Y.

Okey, R.W., *et al.* (1979) Waste-Sludge Treatment in the CPI. *Chem. Eng.*, **86**, 2, 86.

Overcash, M.R., and Pal, D. (1979) *Design of Land Treatment Systems for Industrial Wastes—Theory and Practice.* Ann Arbor Science Publishers, Ann Arbor, Mich.

Parker, D.S., *et al.* (1974) Processing of Combined Physical–Chemical–Biological Sludge. *J. Water Pollut. Control Fed.*, **46**, 2281.

Pavoni, J.L., *et al.* (1973) *Solid Waste Management.* Van Nostrand Reinhold, New York, N.Y.

Process Design Techniques for Industrial Waste Treatment (1974). C.E. Adams and W.W. Eckenfelder (Eds.), EnviroPress, Nashville, Tenn.

Sankey, K.A. (1967) The Problem of Sludge Disposal at the Arnfield Treatment Plant. *J. Inst. Water Eng.*, **21**, 367.

Scher, J.A. (1976) Processing of Waste Oily Sludges. *Proc. Open Forum Manage. Pet. Refinery Wastewaters*, Univ. of Tulsa, Okla.

Sherman, R.J. (1979) Pretreatment of Proteinaceous Food Processing Wastewaters with Lignin Sulfonates. *Food Technol.*, **33**, 6, 50.

Tchobanoglous, G., *et al.* (1977) *Solid Wastes: Engineering Principles and Management Issues.* McGraw–Hill, Inc., New York, N.Y.

U.S. Environmental Protection Agency (1974) *Process Design Manual for Sludge Treatment and Disposal.* Technol. Transfer Div., Washington, D.C.

U.S. Environmental Protection Agency (1975) *Process Design Manual for Suspended Solids Removal.* EPA 625/1-75-003a, Technol. Transfer Div., Washington, D.C.

U.S. Environmental Protection Agency (1979a) *Process Design Manual for Sludge Treatment and Disposal.* Technol. Transfer Div., Washington, D.C.

U.S. Environmental Protection Agency (1979b) *Review of Techniques for Treatment and Disposal of Phosphorus-Laden Chemical Sludges.* Office Res. and Dev., Cincinnati, Ohio.

Viessman, W., Jr., and Hammer, M.J. (1985) *Water Supply and Pollution Control.* Harper & Row, New York, N.Y.

Water Environment Federation (1992) *Design of Municipal Wastewater Treatment Plants.* Manual of Practice No. 8, Alexandria, Va.; Amer. Soc. Civ. Eng. Manual and Report on Engineering Practice No. 76.

Water Pollution Control Federation (1988) *Incineration.* Manual of Practice No. OM-11, Alexandria, Va.

Weber, W. (1972) *Physicochemical Processes for Water Quality Control.* Wiley-Interscience, New York, N.Y.

Chapter 6
Fats, Oil, and Grease Removal

This discussion of pretreatment examines the need for fats, oil, and grease (FOG) control and the effectiveness and economics of control technologies. Past practices and the newer approaches are critically examined so that the industries required to remove such substances can increase the advantages and economic benefits of pretreatment and joint treatment while securing the controls needed to protect effluent quality. Pollution prevention options will be considered for some specific references and examples.

The control of FOG has long been a requirement of many industrial wastewater pretreatment programs at wastewater treatment plants (WWTP). The primary regulation of FOG occurs at the local level to address maintenance problems associated with sanitary sewer blockages and accumulations of floating material in pump station wet wells and treatment tanks. Many sewer-use ordinances and WWTP discharge permits contain limitations regarding FOG either in numerical limits or narrative prohibitions. Fats, oils, and greases are directly or indirectly regulated at the federal level through both general and specific narrative prohibitions, including 40 CFR §403.5(a)(1), which prohibits any pollutant that can cause passthrough or interference with operations at a WWTP. Because some greases can be solid at temperatures found in sewers, 40 CFR §403.5(3) may apply, prohibiting solid or viscous materials that could cause flow obstruction in sewers. Finally, 40 CFR §403.5(b)(6) prohibits the discharge of petroleum oil, nonbiodegradable cutting oil, or products of mineral oil origin that will cause interference or passthrough.

As defined by the U.S. Environmental Protection Agency (U.S. EPA), FOGs from animal or vegetable origin are considered conventional pollutants along with biochemical oxygen demand (BOD), total suspended solids, pH, and fecal coliform (U.S. EPA, 1975). The ability of WWTPs to treat emulsified animal or vegetable FOG has been known for more than 50 years (Mahlie, 1940); however, many WWTPs still place limits of 100 mg/L on the discharge of this material. Extensive research has been presented documenting the compatibility of emulsified animal and vegetable FOG with the biological treatment process used by most WWTPs and should be referred to in requesting variances from the local agencies for the discharge of these materials (Cavagnaro and Kaszubowski, 1988, and ISEO, 1985).

THE NEED FOR FATS, OIL, AND GREASE PRETREATMENT

Pretreatment of FOG is required to avoid interference with the operation of the wastewater collection system and the WWTP. As analytical procedures have improved, discharge limitations have evolved from simple numerical limitations to more sophisticated regulations and local sewer-use ordinances. These local controls include but are not limited to

- Limitations based on total FOG;
- Prohibitions of a number of materials, including free-floating FOG, that can cause obstruction of sewers;
- Prohibition of oil and grease in solid or semisolid form;
- Regulations requiring the removal of floatable oil;

- Limitations on specific kinds of oil, such as petroleum or mineral oils; and
- Prohibition of petroleum FOG.

In many cases, combinations of these provisions are used to provide the WWTP with sufficient authority and control to protect its ability to treat wastewater without interferences.

COLLECTION SYSTEM INTERFERENCES. If FOG globules are large enough to float to the surface in sewers and lift stations, they may obstruct the sewer. These FOG materials should be removed at the industrial facility before discharge. Appropriate industrial pretreatment can prevent FOG from coating, congealing, and accumulating on sewer pipes and pumps and causing obstruction. Effective sewer design and maintenance can also help reduce the problems associated with FOG transport but must be in conjunction with local sewer ordinances (WPCF, 1975; WEF, 1992; WPCF, 1970; and WPCF, 1981).

TREATMENT PLANT INTERFERENCES. Biological Effects. *AEROBIC PROCESSES.* Different kinds of FOG have entirely different responses to aerobic biological treatment. Animal and vegetable oil in a dispersed state is broken down and removed readily in a WWTP. Petroleum-based oils, on the other hand, are degraded slowly by the bacterial process. Such oils, which tend to coat the biological organisms, can prevent or reduce oxygen transfer and degradation of other organics (Groenwold, 1982).

Treatment of dispersed animal and vegetable oil must take into account the treatment capacity of the WWTPs. The conventional activated sludge process is susceptible to shock loads from industrial wastes. This vulnerability may have several causes but is primarily the result of a lack of equalization facilities and an excess of substrate to which the microorganisms are not acclimated (WEF, 1991).

Attempts at aerobic biological treatment of petroleum oil in WWTPs generally has been unsuccessful because of the low rate of biodegradation of the oils. Although some petroleum oil may be removed initially, the process consists primarily of the agglomeration of oil onto clumps of bacteria or other suspended solids and subsequent removal through settling. Unfortunately, treatment of the settled sludge may result in recirculation of the nondegradable oil waste to the plant (see also Chapter 5).

ANAEROBIC PROCESSES. The accumulation of layers of agglomerated semisolid or solid grease, particularly along with hair and other fibrous matter, has caused problems for anaerobic digesters. Liquid oil has also accumulated where supernatant has been drawn from below the surface. Such layers reduce the effective capacity of the digesters. Where the liquid level has been altered, the solid scum layer has placed intolerable stresses on the pipes, supports, and other structural elements within the digesters. Digester shutdown and mechanical removal of the solid matter layer has been necessary at many locations.

In more recent designs, digesters have been provided with means for keeping the contents mixed and heated so as to speed the digestion process and reduce the tendency of scum layers to form. The FOG, when kept well dispersed, is anaerobically degraded and tends to increase the amount of methane gas produced per pound of volatile solids digested (Rudolf, 1944).

Petroleum oil (hydrocarbons), on the other hand, is not degraded to any significant degree in an anaerobic environment. The undegraded oil remains intact and leaves the digester in the supernatant. Where the supernatant is recycled to the incoming wastewater, the oil is also recycled through the system repeatedly.

Physical Problems. High levels of free-floating animal or vegetable FOG can prevent monitoring equipment, such as dissolved oxygen probes, level indicators, pH probes, and other instruments, from operating properly because of clogging or covering of surfaces. This is primarily a problem when a slug of FOG reaches the WWTP. Under normal operating conditions, with routine and periodic maintenance performed on this equipment, FOG is not a problem.

High levels of petroleum FOG can present fire and explosion hazards in a WWTP. Most WWTPs are not designed to protect against or warn operators of explosive atmospheres; thus, it is important to keep these materials out of the sewer.

Floatable FOGs are removed in primary sedimentation. The proportions removed in the skimmings and waste sludge can be substantial. Pumping skimmings and sludge with high FOG concentrations has caused trouble in many WWTPs. The material congeals on the walls of the WWTP piping and often obstructs flow completely.

Fats, oils, and greases have also been found to cause problems with dewatering of raw and digested primary sludge. For example, if present in large amounts, free-floating FOGs may contribute to the plugging of filters.

Passthrough. Because petroleum FOG degrades slowly in WWTPs, if sufficient quantities exist in the influent, it can pass through the treatment plant (Groenwold, 1982). Passthrough of a petroleum FOG can cause a violation of the WWTP's discharge permit because of a violation in water quality standards for floating materials or an inability to meet any other monitoring requirement, including biotoxicity. Animal or vegetable FOG in excess quantities in the WWTP's influent can also pass through.

CHARACTERISTICS OF FATS, OIL, AND GREASE

Fats, oil, and grease may be present in water either as an emulsion or free floating. Most heavy oils and greasy materials are insoluble in water but may be emulsified by contact with detergents, alkalies, other chemicals, or mechanical means such as high-speed pumping. Unexpected FOG test results can occur when detergents are present in a sample. If detergents are present at low concentrations (below the critical micelle concentration), they will not be recovered by any of the analytical methods. (A micelle is an electrically charged colloidal particle usually organic in nature, composed of aggregates of large molecules, for example, in soaps and surfactants. The term is applied to the casein complex in milk. The critical micelle concentration is the concentration at which micelles begin to form in the bulk liquid.) However, when surfactants are present in high concentrations, a portion will be recovered by the analytical procedures and, thus, be classified as FOG.

The term FOG includes materials of vegetable, animal, and mineral origin. Mineral oils include petroleum, hydrocarbon, or nonpolar FOG.

Fats are found in both plants and animals and are among the most important components of the human diet. Fats are the glycerol esters of fatty acids, also called triglycerides. They are classified by chemists according to their average molecular weight and degree of unsaturation. Fats are mixtures of various triglycerides rather than pure individual compounds and have low and generally nonspecific melting points. A predominance of unsaturated fatty acid components lowers the melting point of mixtures. When a triglyceride is liquid at room temperature, it may be referred to as an oil. Common commercial edible oils are cottonseed, palm, olive, corn, and soybean. Common fats are lard, tallow, and butter fat.

Soap is formed when animal fat or vegetable oil is boiled with sodium hydroxide to produce glycerine and sodium salts of fatty acids. Soaps are included in the analysis typically performed for FOG; that is, they are extracted from an acidified sample in the analytical procedures.

Waxes are the monohydroxylic alcohol esters of fatty acids. Waxes differ from fats and oils in being esters of long-chain normal primary alcohols. At room temperature, waxes are much harder than fats, and their biological function is generally that of a protective coating or a structural material, such as beeswax. Natural waxes contain free acids, free alcohols, and other hydrocarbons. Waxes are included in typical FOG analyses.

Grease is a general classification for grouping such materials as fats, oils, waxes, and soaps according to their effect on wastewater collection and treatment systems or their physical (semisolid) forms.

ANALYTICAL PROCEDURES

Knowledge of the variety of compounds that can be included in the analytical results obtained from testing for FOG is necessary to plan and implement an adequate pretreatment program. Analyses required for compliance evaluation may not be sufficient for treatment system design or operation. The degree of control needed, the treatment methodology to be applied, and the resulting environmental impact vary for the different types of FOG materials. *Standard Methods* defines FOG as groups of substances with similar physical characteristics that are "determined quantitatively on the basis of their common solubility in trichlorotrifluoroethane. It may include hydrocarbons, fatty acids, soaps, fats, waxes, oil, and any other material that is extracted by the solvent from an acidified sample and that is not volatilized during the test" (APHA, 1992).

In recent years, Freon™ (trichlorotrifluoroethane) has been selected as the analytical solvent, mainly for safety reasons. However, interferences can be associated with trichlorotrifluoroethane use because other organic substances, if present, may also be dissolved. Unfortunately, there is no known solvent that will dissolve only FOG. New methods under development will eliminate the use of fluorocarbon-based solvents; *n*-hexane and methylene chloride are likely replacements.

Instrumental techniques such as gas or liquid chromatography also may be used for FOG determinations. However, this methodology typically cannot be used routinely because of the cost and complexity of sample preparation and instrumentation.

SAMPLING. Obtaining representative samples for analysis of FOG from wastewater and effluent may be difficult. Fats, oil, and grease can adhere to various parts of automatic sampling equipment during collection of composite samples and, thereby, adversely affect representative analysis. Fats, oil, and grease samples, therefore, should be collected below the wastewater surface by grab-sampling techniques. A widemouth glass bottle should be used for collecting the sample. Collection never should be made from the overflow or immediately downstream of a weir, or sloughing of FOG accumulations caused by the overflow structure or weir could result (U.S. EPA, 1973a).

When analysis cannot be performed immediately, the sample should be preserved with a few drops of concentrated sulfuric acid using standardized preservation procedures (APHA, 1992). Separate samples should be collected for each FOG determination. Subdividing a total composite sample in the laboratory should be avoided because of the tendency of FOG to adhere to the sides of the sampling container and lid, causing substantial analytical error. Samples should be collected and stored in sample containers provided by the analytical laboratory.

ANALYTICAL METHODS. Several recognized standard methods exist for determining FOG. When results are presented, the procedure selected to determine the FOG fraction of a wastewater stream, as well as the solvent used, should be stated. The determination should also be completed within a few hours after collection because standing may cause a gradual weight loss resulting from volatilization of lighter constituents (APHA, 1992, and U.S. EPA, 1973a).

According to pretreatment regulations, WWTP sampling programs must specify that only approved U.S. Environmental Protection Agency (U.S. EPA) analytical methods be used to determine compliance with a standard. A listing of these methods can be found in the Code of Federal Regulations, 40 CFR §136.3. As of this writing, U.S. EPA method 413.1 (partition gravimetric) and *Standard Methods*, 16th Ed., 503A, are the only approved methods for FOG analysis.

In addition to the approved method, three other FOG determination methods are currently being used for nonregulatory purposes. These include U.S. EPA method 413.2 (oil and grease, total, recoverable, spectrophotometric, infrared), U.S. EPA method 418.1 (petroleum hydrocarbons, total recoverable, spectrophotometric, infrared), and *Standard Methods* (APHA, 1992) 2530 C (trichlorotrifluoroethane—soluble oil and grease).

Total Fats, Oil, and Grease. Methods 413.1 and 413.2 require acidifying a sample to pH < 2. A liquid–liquid extraction is then performed using an organic solvent to extract the FOG from the sample. Once the FOG is contained in the organic solvent, it can be "measured" either by drying at 70°C (158°F) and weighing or by comparison of the infrared absorbance of the extract with standards.

U.S. EPA methods 413.1 and 413.2 differ only in the detection method for the extracted FOG materials. Method 413.1 uses gravity to measure the total amount of material extracted from the sample. This makes this method much less specific than U.S. EPA method 413.2, which uses a standard calibration curve and infrared absorbance to measure the amount of FOG materials extracted from the sample. Method 413.1 uses heat to dry the extract; thus, the lighter fraction (those with boiling points < 70°C) of FOG may be lost, resulting in a negative interference.

Total Petroleum Hydrocarbon. Method 418.1 consists of acidifying a sample to pH < 2. A liquid–liquid extraction is then performed using an organic solvent to extract the FOG from the sample. The solvent containing the extracted FOG is filtered through silica gel to remove the polar organic fraction (assumed to be animal/vegetable FOG). The remaining extracted FOG is measured using infrared absorbance by comparison with a calibration standard.

Floatable Fats, Oil, and Grease. The determination of floatable FOG is significant in that a gravity separator can only remove this type of FOG; thus, this test is useful for determining removal efficiencies. Also, floatable FOG is more likely to cause sewer obstruction, so this test can be used to monitor indirectly sewer obstruction potential.

The floatable FOG test is conducted by allowing a 1-L grab sample of the waste stream to settle for 30 minutes in a 300-mm (11.81-in.) tall by 70-mm (2.76-in.) inside diameter glass tube. After settling, the water is withdrawn from the bottom of the settling tube. The remaining "floatable" material retained in the tube is acidified to pH < 2, then the FOG is extracted with an organic solvent as in the other FOG methods. The sample is dried of solvent at 70°C (158°F) and weighed.

Combined U.S. Environmental Protection Agency Methods 413.2 and 418.1. By using a combination of methods 413.2 and 418.1, a better understanding for the nature of FOG in the sample can be obtained. Because method 413.2 measures total FOG and method 418.1 measures total petroleum hydrocarbon FOG, the difference between the two test results can be used to determine the amount of polar FOG in the sample (normally assumed to be animal/vegetable FOG). The use of method 413.2 is preferred in this calculation so that both results are determined using the same detection method (infrared absorbance).

SOURCES OF FATS, OIL, AND GREASE

In domestic wastewater, FOG may average 30 to 50 mg/L and be as much as 20% of the organic matter measured as BOD. Wastewater from communities with industrial contributors typically have higher concentrations of FOG. Major commercial and industrial contributors of FOG to WWTPs are listed in Table 6.1.

FOOD-PROCESSING INDUSTRY. Within the food industry are a variety of potential sources of FOG, including the processing and rendering of meat, cooking and processing of vegetables, production of edible oils, and processing of nuts and seeds. Fats, oil, and grease from these processes are highly biodegradable; thus, only the floatable portion of these wastes requires pretreatment before discharge to a WWTP. Gravity separation and pH adjustment are the most common types of pretreatment in this industry (ISEO, 1985).

METALWORKING INDUSTRY. Metal-cutting fluids from most metal machining operations (also called lubricants, coolants, and cutting oils) contain

Table 6.1 Industries that are major contributors of fats, oil, and grease to wastewater treatment plants.

Industry	Type of FOG
Vegetable oil refining	Vegetable
Soap manufacturing	Vegetable and animal
Milk processing	Animal
Dairy products, including cheese	Animal
Rendering	Animal
Slaughterhouse and meat packing	Animal
Candy manufacturing	Vegetable
Food preparation	Animal and vegetable
Eating establishments	Animal and vegetable
Laundry	Animal, vegetable, and petroleum
Metal machining	Petroleum
Metal rolling	Petroleum
Tanneries	Animal and vegetable
Wool processors	Animal
Petroleum refineries	Petroleum
Organic chemical manufacturing	Petroleum, animal, and vegetable

mineral oils. These oils are derived from petroleum and coal. Wastewater containing these oils can originate from a variety of sources, including machine shops, auto garages, car washes, gasoline stations, laundries, refineries, steel mills, tanneries, wool scouring, drop-forge plants, and stormwater. Grease is a form of mineral oil that is formed by heating a mixture of lubricating oil and soap.

Workpieces processed in a metal finishing plant are frequently covered with FOG from machining operations or from application for surface protection during storage and shipping. Fats, oil, and grease typically is removed by using organic solvents and/or inorganic alkaline cleaning solutions. Effluents usually are contaminated with these water-immiscible solvents as a result of dragover or batch dumping of the cleaning media.

The solvents used in vapor or immersion types of degreasing (for example, nonflammable chlorinated hydrocarbons or solvents such as kerosene used in emulsion cleaners) can form water emulsions or a floating film that may be toxic to microorganisms found in WWTPs. Additionally, these organic contaminants in effluents may be flammable or liberate toxic gases, which would prohibit their discharge to public sewer systems (U.S. EPA, 1973b).

Although organic and a potential source of energy for bacteria, FOG of mineral origin is generally not compatible with WWTP treatment processes. Modern lubricating oils and greases are being formulated with limited or no mineral oil content. These are known as high water content fluids (HWCF), which contain a maximum of 3 to 15% mineral oil emulsified in water. Synthetic HWCFs contain no mineral oil, the lubricating effect being provided by other water-soluble organics (Lawrence, 1983).

PETROLEUM INDUSTRY. Wastes from petroleum refineries include free and emulsified oil from leaks, spills, tank draw-off, and similar sources; emulsions incident to chemical treatment, oil-containing condensate, waters from distillate separators, and tank draw-off; and oil-containing alkaline and acidic wastes and sludge. The combined refinery wastewater discharge may contain crude oil, various crude oil fractions, or suspended material coated with oil, soaps, and waxy emulsions (Nemerow, 1971).

Also included in the petroleum FOG category are light hydrocarbons such as gasoline and jet fuel; heavy hydrocarbon fuels and tars such as crude oils, diesel oils, and asphalt; lubricants; and cutting fluids.

*P*RETREATMENT TECHNIQUES

Free FOG—that which is not emulsified—presents no serious problem with respect to its removal from water because it will float to the surface and agglomerate. The FOG then can be mechanically skimmed or lifted from the surface. Emulsified FOG, however, because it stays in suspension, causes severe separation problems. Normally, FOG should be recovered as close as possible to the source, thereby reducing the size of treatment units and minimizing the locations at which grease can accumulate. To prevent emulsification of free FOG through agitation, pumping of the wastewater should be avoided wherever possible. Diluting the FOG-containing wastes with other non-FOG-bearing waste also should be avoided so that maximum detention time may be achieved in the removal system.

Before any pretreatment process for emulsified FOG removal is selected or installed, characterization and treatability studies should be performed so that an effective and efficient FOG removal process can be designed. Numerous mechanical devices exist today that are designed to remove FOG from wastewater discharges. Such equipment may or may not be applicable to a particular FOG removal opportunity. Operation and maintenance savings from a properly and efficiently designed pretreatment system will easily offset the cost of any up-front studies.

Fats, oil, and grease treatment may be categorized into a first and second stage. First-stage treatment can be used to separate free floatable FOG from water and emulsified substances. The treatment process, which involves gravity separation of the FOG material, is equally effective in removing fats, greases, and nonemulsified oils. In some instances, gravity separation can be enhanced with coalescing media or parallel plate designs.

Second-stage treatment involves emulsion breaking and removal of emulsified FOG. Emulsion breaking technologies include heating, distillation, chemical treatment plus centrifugation, chemical treatment plus precoat filtration, and filtration. Recently, ultrafiltration has been successfully used in cutting oil and fatty acid recovery systems. The most common second-stage FOG removal installation consists of simple gravity separation followed by chemical addition (alum or ferric chloride), flocculation, then a dissolved air flotation (DAF) unit.

GRAVITY SEPARATION. The simplest form of first-stage treatment is the oil interceptor, or grease trap, used in small establishments like restaurants, hotels, and service stations. The grease trap is a receptacle designed to collect and retain FOG substances normally found in kitchen or similar wastes. It typically is installed in the drainage system between the building sewer and the point of production of the wastes. Fat, oil, and grease interceptors ideally are placed at locations that are readily accessible for cleaning and maintenance. Interceptors can be useful in industrial applications where small intermittent flows are involved and the volume of FOG is small.

For industrial wastes from rendering plants, food processing facilities, and oil refineries, a larger gravity separator with mechanical float removal and waste oil storage is required. In the food industry, the waste oil from the gravity separator may be recovered for reuse or sale. The standard criteria used for designing gravity separators have been developed and published by the American Petroleum Institute (API, 1959). Large separators may be operated as batch tanks or continuous-flow-through basins, depending on the volume and type of waste to be treated. Additional guidance on separator sizing has been provided by Thomson (1973). A study of gravity separator operating data showed that the level of floatable FOG remaining after treatment in a properly designed and operated gravity separation system does not result in sewer obstruction (McDermott and Yee, 1986).

The goal in designing a gravity separator is to provide sufficient area and quiescent conditions for a long enough period of time to allow the FOG to float out of the mixture. Any flow disturbances will prevent this goal from being met. Changes in influent flow rate can affect removal efficiency. Control of upstream sources with surge tanks may be required to meet this goal. If steady flow is not anticipated because of other considerations, the trap should be oversized based on the peak sustained flow rate.

Gravity separators range in size from small packaged restaurant units to huge industrial product recovery systems. When designing a gravity separator or fat trap, it is important to consider the hydraulic capacity of the unit, the flow configuration, and the ease and convenience with which the trap may be cleaned. The distance between the inlet and the outlet of the trap should be sufficient to allow gravity differential separation so that grease will not escape through the outlet. This can be calculated by making some assumptions about the oil or grease to be removed and the use of Stokes' law for the terminal rise velocity of a particle. Assuming the smallest oil globule to be removed is 0.015 cm (0.005 9 in.), the rise rate can be calculated as follows (API, 1959):

$$V_t = 1.224 \times 10^{-2} \, [(S_w - S_o)/\mu] \qquad (6.1)$$

Where

V_t	=	rise rate of oil globules of 0.15 mm (0.005 9 in.) or more, mm/s;
S_w	=	specific gravity of waste at design temperature;
S_o	=	specific gravity of oil in wastewater at design temperature; and
μ	=	absolute viscosity of waste at design temperature, poises (Ns/m).

With gravity separation, the terminal rise rate of a settling particle can be converted to an overflow rate by converting units. Thus, V_t, after conversion to units of gallons per square feet per day, can be used to design a fat trap using the following relationship:

$$V_t = d/t = (d/LBd)/Q_m = Q_m/LB = \text{Overflow rate} \qquad (6.2)$$

Where

d	=	depth of water in separator, m;
t	=	retention time in separator, seconds;
L	=	length of separator, m;
B	=	width of separator, m;
Q_m	=	waste flow rate, m^3/s; and
V_t	=	overflow rate, m^3/m^2· d, which is equivalent to the rise rate, m/s, of the smallest particle to be removed.

From the relationship between flow velocity and volume, the overflow rate can be used to estimate the surface area required for effective separation as follows:

$$A_H = f(Q_m/V_t) \qquad (6.3)$$

Where

A_H	=	minimum surface area for gravity settling, m^2 (this is normally multiplied by a correction factor, f) and
f	=	flow short-circuiting and turbulence factor, usually ranges from 1.2 to 1.8; the higher the (V_H/V_t), the larger the value of f.

The following are rules of thumb for good design:

- A minimum cross-sectional area should be used, with a maximum ratio of oil rise rate to horizontal flow velocity of 15, and a maximum horizontal velocity of 15.4 mm/s (3 ft/min);
- A minimum depth of 1.22 m (4 ft) and a maximum depth of 2.44 m (8 ft); and
- A maximum depth-to-width ratio of 0.5 and minimum of 0.3.

Flow control baffles are essential, and flow control fittings also may be necessary on the inlet side of smaller traps to protect against overloading caused by sudden wastewater surges. A means to retain free-floating FOG is required, such as an effluent baffle. Internal baffling should not be used because it increases turbulence within the tank and may reduce effective surface area.

In some cases, where the waste greases are hot, placing the trap as close as possible to the source is particularly advantageous because the grease remains liquid and can be skimmed and pumped easily. In other instances, it is easier and more efficient to remove cooled greases because they are solid and can be removed mechanically. However, as grease builds up, the working volume of the unit will be decreased, potentially decreasing the efficiency of the trap. With units that keep the grease in a liquid form, it can be removed continually before it cools and solidifies, thus maintaining a constant working volume. Small FOG interceptors should have FOG-retaining capacity in pounds equal to at least twice the flow rate in gallons per minute (litres per second) (PDI, 1965). Slightly oversized interceptors generally are more economical because they require less maintenance. No matter what size the FOG interceptor, the unit must be operated properly and cleaned regularly so that grease will not escape.

Coalescing Gravity Separators. Commercial devices are available for removing oil from wastewater by passing it through a medium with a large surface area. Typically, a plastic medium is used with a high affinity for oil. The combination of agglomeration resulting from flow path disruption and the impaction with the media causes the oil globules to grow in size. Because larger oil globules are removed easier by gravity, treatment efficiency for a given surface area increases. The medium should be investigated to ensure that it will not render the recovered oil unusable. (Food and Drug Administration-approved materials must be used if captured oil is intended for human consumption or animal feed.) Also, many animal and vegetable oils are more polar than petroleum oils and may adhere to plastic media, leading to fouling. This type of separator works best with light FOG loadings and requires more maintenance than simple gravity separators because of the need to keep the medium free of blockages.

Chemically Enhanced Separation. Chemical treatment of an emulsion typically is directed toward destabilizing the dispersed oil or destroying any emulsifying agents present. The process consists mainly of rapid mixing of a coagulant chemical with the wastewater, followed by flocculation and flotation or another physical separation technique. Various chemicals have been used in this demulsifying process, including coagulating salts, acids, salts plus heat, and organic polyelectrolytes. Coagulation using aluminum or iron salts has been effective for demulsifying oily wastes, but the precipitated wastewater solids is difficult to dewater, and the volume of sludge generated and requiring disposal creates an additional problem. When acids are used to break emulsions, more expense is involved because the resultant clarified wastewater must be neutralized before discharge to a WWTP. The addition of large quantities of an inorganic salt may also create subsequent pollution problems by significantly increasing the dissolved solids content of the wastewater discharge. Some organic polyelectrolytes that are extremely effective in demulsifying FOG are also quite expensive.

The use of chemicals to enhance gravity separation depends largely on the type of materials involved and the type and stability of the emulsion. Batch settling tests should be conducted at the bench before any field work. A traditional "jar test" is the best means to identify both the type and dose of chemical required to remove FOG materials. Control of pH is also important in emulsion breaking and should be increased during testing.

If the chemical character of the waste stream changes, periodic "jar tests" may be required to determine adjustments to the chemical doses. If the waste character changes rapidly and frequently without warning, chemical addition systems should be used with caution.

When the floc particle created through chemical treatment is not readily separated by gravity, DAF or centrifugation can be used to aid the separation. However, if chemical addition does not produce a stable floc, these separation techniques will not work effectively.

DISSOLVED AIR FLOTATION. After chemical additions succeed in breaking an emulsion, the FOG typically is concentrated and removed by a physical process. The most widely used physical process, DAF (Figure 6.1), consists of adding air to the waste stream under pressure with a pump, then releasing the air-charged stream to atmospheric pressure in a tank. The oil and small solids cling to the minute air bubbles and float to the surface where they are skimmed off (see Chapter 5 section Flotation).

With DAF, two schemes are used: recycle pressurization and direct pressurization. Considerations that control the design and operation specifications include air-to-solids (oil) ratio (A:S); operating pressure of the direct or recycle flows; pressurized flow rate; influent flow rate; and rise velocity of the air–oil mixture. The following two equations are used for determining the A:S ratio.

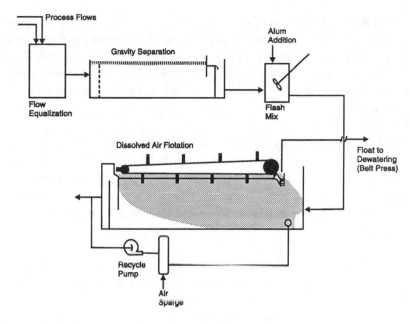

Figure 6.1 Typical dissolved air flotation system.

$$\text{Recycle pressurization A:S} = [1.3A_s(fP - 1)R]/QS_a \qquad (6.4)$$

$$\text{Direct pressurization A:S} = [1.3A_s(fP - 1)]/S_a \qquad (6.5)$$

Where

S_a	=	influent suspended solid or FOG concentration, mg/L;
A_s	=	air saturation, cm³/L;
R	=	pressurized flow rate, m³/d;
Q	=	influent flow rate, m³/d;
P	=	operating pressure (absolute) measured, atm ($P = [p$ {gauge pressure, kPa} + 101.35]/101.35); and
f	=	fraction of air dissolved at pressure P, typically 0.8.

Particle rise rate and recycle rates are also important design considerations. More detail about DAF can be found in Chapter 5.

The proper A:S ratio for the successful use of air flotation to remove FOG depends on the concentration and characteristics of the FOG in the influent. A flotation cell can be used in the laboratory to determine the optimal A:S ratio. Typically, a ratio between 0.005 and 0.06 is desirable for efficient flotation. Poor capture generally results when the ratio falls below 0.01. Variation in influent concentration will alter the A:S ratio. The operator must readjust this ratio by changing the recycle flow rate, the operating pressure, or both, to keep the process at its highest efficiency.

CENTRIFUGATION. When floc particles created with chemical treatment do not separate well, a centrifuge can be used to make the separation process more efficient. Centrifuges require more maintenance and energy than do other types of separators but can be used when space is at a premium. Pilot testing is required to determine if good separation can be achieved using a centrifuge.

FILTRATION. Filtration has been used with some success, for instance with high-rate sand and precoat diatomaceous earth filters. The addition of coagulating chemicals is necessary to increase the removal of oil beyond that achieved by gravity settling. The coagulants create larger particles and enmesh the oil particles, thereby forming agglomerations large enough to be removed.

BIOLOGICAL REMOVAL. For many FOG materials not removed by gravity or DAF and still unacceptable to the WWTP, biological pretreatment may be required. A WWTP designed to treat typical components of domestic wastewater may not provide the hydraulic and solids retention times necessary to allow high volumes of animal and vegetable or petroleum FOG to be removed effectively. This does not mean that biological treatment is inappropriate. If the FOG material can be degraded biologically, this form of treatment may be advisable. Both aerobic and anaerobic processes have been used to treat FOG materials (U.S. EPA, 1973a, and Nemerow, 1971). Sludge disposal must be considered when deciding to use biological treatment (see Chapters 3 and 9 for more details).

ULTRAFILTRATION. Ultrafiltration is currently in use for the recovery of fatty acids (C_{10}–C_{22}) from the soap manufacturing process (Tollett and Shedroff, 1990). The system relies on the ability of fatty acid to form stable micelles in water. These micelles are larger than the filter pore size and effectively filtered out of the wastewater. Ultrafiltration breaks the oil–water emulsion but can only concentrate the FOG in water, not remove it completely. A gravity coalescing filter is used to remove the fatty acid from the concentrate stream. The removal of the FOG depends on the fatty acids becoming insoluble as their concentration increases; thus, they can be removed by the coalescing gravity separator.

Ultrafiltration is also in use similarly for the recovery of machine oils in metalworking shops. The economics of these systems depend greatly on the value of the recovered material. Micellular-enhanced ultrafiltration of other oily materials currently is under development. The major advantage over chemical addition systems is that the resulting recovered material is free of chemical additives and, thus, can be recycled or recovered, with no sludge that requires disposal being produced.

OPTIONS FOR RECOVERED OIL AND GREASE

The oil recovered by gravity settling of industrial wastes from industries such as edible oil refining, soap making, rendering, and meat processing can be salvaged and used in animal foods. Many restaurants collect the spent FOG from frying vessels and sell it to rendering plants, where it is purified and subsequently sold for industrial or animal feed use. Oil skimmed from gravity separators may be included, or it may be discarded with other solid waste and refuse from the restaurant.

Petroleum hydrocarbons with low water content can be used in refinery feed stock or for fuel. The increased price of petroleum has reduced the volume of such oil and waste material for disposal. Waste motor oil from service stations and waste lubrication oil from certain industries can be collected by waste haulers and profitably sold to waste oil refiners.

REUSE. There may be other uses for collected waste FOG, depending on the type and percentage of oil, fat, or waxes. The potential marketplace for recycled oil should be surveyed through such media as industrial waste exchanges now being formed throughout the country. The sale of FOG collected from pretreatment systems may help offset the associated operating costs.

Three industrial processes that present FOG reuse options are drilling mud manufacturing, ore flotation, and asphalt manufacturing. Many drilling mud manufacturers can use a variety of FOG materials as a raw material or product ingredient. Most types of FOG are acceptable. Suitability for this use must be determined on a case-by-case basis.

Ore flotation is the process by which mineral ores are recovered after acid extraction. The process uses fatty acids produced from animal and vegetable fats and oils to make the metal ions capable of floating in a DAF-like system. Fats, oil, and grease consisting of fatty acids may be usable in the ore-flotation industry. The FOG may require processing before use. Suitability for this use must be determined on a case-by-case basis.

Some asphalt manufacturers can use certain FOG material to aid in emulsification of other materials used in the production of asphalt. Suitability for this use must be determined on a case-by-case basis.

Polar FOG of animal or vegetable origin can be recovered and sold to be used in making animal feed. A market for these materials has developed over the years.

Animal and vegetable oils separated by coagulation with chemicals may be difficult to use as animal feed. Because the recovered material must be treated to remove those coagulants (most of which are unacceptable in animal feed), the cost of chemical removal generally exceeds the value of the product.

Finally, a number of companies are in the business of collecting "trap stock" FOG streams (in the form of DAF float) and reselling it for the recovery of the oil to be used in animal feed, tallow, and cosmetics.

RECYCLE. In many cases, the most economical alternative is to develop methods to return recovered FOG materials back to the process that generates them. This requires that the waste oil collection system be separated from other wastewater processes, a stipulation that should be considered when designing new facilities as a means to prevent pollution. For example, many chemical plants use water to wash impurities from finished product or intermediates. This water may contain valuable product as a result of process control limitations. The concentration of finished product in the wash water may be low, but over time a large amount of material could be lost. For products that float, gravity separation is widely used for recovery. The recovered material is reprocessed on site. This is typically installed as part of the process.

The ability to recycle, reuse, incinerate, or otherwise dispose of the recovered FOG materials is regulated by the Resource Conservation and Recovery Act (RCRA). Consult with this regulation before making any disposal option decisions. The ability to sell a chemical substance, even a recycled substance, will depend on its inventory status for all U.S. uses not excluded for the Toxic Substances Control Act. For example, a waste from the food industry may require inventory listing to be sold to the detergent, lubricant, or other non-U.S. Food and Drug Administration regulated industry.

LANDFILL. Systems that use chemical addition to break emulsions typically generate sludge. The sludge can be dewatered using a filter press (or other method; see Chapter 5), then landfilled. Where possible, generation of this sludge should be avoided. The long-term costs associated with landfilling sludge is hard to predict because of rapidly changing space availability and disposal economics.

Disposal of solidified FOG materials may be appropriate if no other economically viable recovery options exist. Landfilling should be used only as a last resort disposal method.

INCINERATION. Because the majority of FOG materials are organic, they can be incinerated. In some cases, dry FOG recovered from gravity separation has a good heat value and may be inexpensive to incinerate because of its value as fuel. Sludge generated by chemical addition systems may not be incinerable because of salt or metal content.

The ability to use recovered FOG materials for their heat value is regulated by RCRA. Consult with this regulation before making any disposal option decisions regarding recovery of heat or incineration (see Chapter 5).

TRANSPORT TO ANAEROBIC DIGESTION. Collection of concentrated FOG emulsion waste for transport to the WWTP, where it is introduced directly into an anaerobic digester, is another possibility. This eliminates dilution in the collection system, removal in the WWTP's primary clarifiers, and other processing steps. Introduction of FOG should be at a controlled rate to reduce the likelihood of digester upset.

*R*EFERENCES

American Public Health Association (1992) *Standard Methods for the Examination of Water and Wastewater.* 18th Ed., Washington, D.C.

American Petroleum Institute (1959) *Manual of Disposal of Refiner Wastes.* Vol. 1, New York, N.Y.

Cavagnaro, P.V., and Kaszubowski, K.E. (1988) Pretreatment Limits for Fats, Oils, and Grease. *Proc. 43rd Ind. Waste Conf., Purdue Univ.*, West Lafayette, Ind., 777.

Groenwold, J.C. (1982) Comparison of BOD Relationships for Typical Edible and Petroleum Oils. *J. Water Pollut. Control Fed.*, **54**, 4.

Institute of Shortening and Edible Oils, Inc. (1985) *Treatment of Wastewaters from Food Oil Processing Plants in Municipal Facilities.* Washington, D.C.

Lawrence, P.R. (1983) Oily Wastewater Treatment and the Impact of High Water Content Synthetic Fluids at the Ford Motor Company. *Proc. 38th Ind. Waste Conf., Purdue Univ.*, West Lafayette, Ind., 29.

Mahlie, W.S. (1940) Oil and Grease in Sewage. *Sew. Works J.*, **12**, 3, 527.

McDermott, G.N., and Yee, N.S. (1986) Practical Limitations of Oil in a Pretreatment Program. Paper presented at the Tenn. Water Pollut. Control Assoc. Meeting.

Nemerow, N.L. (1971) *Liquid Waste of Industry; Theories, Practices and Treatment.* Addison-Wesley Publishing Co., Reading, Mass.

Plumbing and Drainage Institute (1965) *Testing and Rating Procedure for Grease Interceptors.* Plumbing and Drainage Stand. PDI-G101, Oak Park, Ill.

Rudolf, W. (1944) Decomposition of Grease During Digestion, Its Effects on Gas Production and Fuel Value of Sludges. *Sew. Works J.*, **16**, 6, 1125.

Thomson, S. J. (1973) Report of Investigation on Gravity-Type Oil Water Separators. *Proc. 28th Purdue Ind. Waste Conf., Purdue Univ.*, West Lafayette, Ind., 558.

Tollett, R.M., and Shedroff, S.A. (1990) Development of Ultrafiltration to Control FOG. *Proc. New England Environ. Expo*, 883.

U.S. Environmental Protection Agency (1973a) *Handbook for Monitoring Industrial Wastewater.* Technol. Transfer, Cincinnati, Ohio.

U.S. Environmental Protection Agency (1973b) *Waste Treatment: Upgrading Metal Finishing Facilities to Reduce Pollution.* Technol. Transfer Publ.

U.S. Environmental Protection Agency (1975) *Treatability of Oil and Grease Discharged to Publicly Owned Treatment Works.* Effluent Guidelines Div., EPA/440/1-75-066, Washington, D.C.

Water Environment Federation (1991) *Operation of Municipal Wastewater Treatment Plants.* Manual of Practice No. 11, Alexandria, Va.

Water Environment Federation (1992) *Wastewater Collection Systems Management.* Manual of Practice No. 7, Alexandria, Va.

Water Pollution Control Federation (1970) *Design and Construction of Sanitary and Storm Sewers.* Manual of Practice No. 9, Washington, D.C.

Water Pollution Control Federation (1975) *Regulation of Sewer Use.* Manual of Practice No. 3, Washington, D.C.

Water Pollution Control Federation (1981) *Gravity Sanitary Sewer Design and Construction.* Manual of Practice No. FD-5, Washington, D.C.; Am. Soc. Civ. Eng. Manuals and Reports on Engineering Practice No. 60, New York, N.Y.

Chapter 7
Neutralization

Neither excessively acidic nor alkaline wastewater effluent is acceptable for direct discharge to surface waters. Thus, a common chemical treatment of wastewater is pH adjustment. The process of adjusting pH to approximately 7.0 is called *neutralization*. A pH range of 5 to 10 may be acceptable for pretreatment purposes, but tighter limits such as 6.0 to 8.5 are more common for direct discharge.

Adjustment of wastewater pH is often necessary for adequate treatment at municipal or industrial wastewater treatment plants (WWTP). The pH correction of strongly acidic or alkaline wastewater is required before discharge of the waste to a sewer or before treatment by biological or physicochemical means (Okey and Chen, 1978; Baker, 1974; and Nemerow, 1971). Strongly acidic or alkaline wastes may seriously impair the performance of WWTP unit processes and may be destructive to treatment and collection facilities.

The range of pH required for effective biological wastewater treatment is fairly narrow. Generally, the optimum pH for bacterial growth lies between pH 6.5 and 8.5. Optimum growth of nitrifying bacteria, however, is typically observed in the pH range of 8.0 to 8.5. Biological conversion of ammonia (NH_3) to nitrate (NO_3^-) is called *nitrification* and produces a more acidic state. If sufficient initial alkalinity is not present to neutralize the acidity produced, the pH of this system will drop and nitrification may be inhibited.

Wastewater discharged at low pH into a sewerage system can trigger adverse chemical reactions in the sewers. When cyanide ion in wastewater comes into contact with low-pH waste, the combination creates hydrogen cyanide gas, which is highly toxic. Sulfides present in wastewater may also evolve hydrogen sulfide gas when the pH of the wastewater is low. Both of these gases can harm sewer maintenance workers and may also be toxic to treatment plant workers in enclosed areas. In addition, hydrogen sulfide gas can be oxidized biologically to form sulfuric acid, which will corrode the crown of unprotected gravity flow sewers.

The characteristics of wastewater to be neutralized vary from plant to plant and within a facility itself. The economics of pH correction frequently require a tradeoff between complete neutralization of the waste (adjustment to pH 7.0) and the chemical or dilution costs to achieve pH control within allowable discharge limits. In addition, selection of the pH adjustment method and facility design should take into consideration the potentially deleterious reactions that may take place in a collection system as a result of chemical additions. Thus, the process of neutralization is more than just mixing the computed amount of neutralizing agent with wastewater. A successful process design includes examination of a number of elements that influence the design. These elements should be examined in a planned sequence before design so that the resulting facility will be efficient and economical in both capital and operating costs. Elements of concern and a successful sequence are given below:

1. Wastewater characteristics (alkalinity or acidity) and variability;
2. Wastewater flow rate and variability;
3. Discharge criteria;
4. Performance and cost analysis of alternative neutralizing agents;
5. Selection of neutralizing agent and reaction effects;
6. Selection of neutralizing process;
7. Process design and control strategy; and
8. Facility design, construction, start-up, and operator training.

The design of neutralization facilities should be conducted after waste reduction and resource recovery opportunities have been fully evaluated. Furthermore, neutralization opportunities through blending and equalization of waste streams should also be evaluated within a single industrial establishment or between neighboring industrial plants.

Table 7.1 lists several industries that produce acid and alkaline wastewaters. The table is for illustrative purposes only and not intended to be a complete listing.

*T*ERMS AND DEFINITIONS

The following subsections present the fundamental concepts of pH, pOH, acidity, alkalinity, and buffer capacity. Additional information, examples, and problems and solutions are presented elsewhere (see APHA, 1992).

Table 7.1   Industries that produce acid and/or alkaline wastewater.

Industry	Type of wastewater generated
Aluminum	Acid
Brass and copper	Acid and alkaline
Brewery and distillery	Acid and alkaline
Cannery	Acid and alkaline
Carbonated beverages	Alkaline
Chemical plants	Acid and alkaline
Chemical softening wastes	Alkaline
Coal mining drainage	Acid
Coffee	Acid
Explosives	Acid
Glue	Acid and alkaline
Iron and steel processing	Acid
Laundry, commercial	Alkaline
Leather tanning	Acid and Alkaline
Metal pickling	Acid
Oil refinery	Acid and alkaline
Pesticide	Acid
Pharmaceutical	Acid and alkaline
Phosphate	Acid
Powerhouse blowdowns and cleanings	Acid and alkaline
Pulp and paper	Acid and alkaline
Regeneration of ion exchange and demineralizers	Acid and alkaline
Rubber	Acid and alkaline
Textile	Acid and alkaline

pH AND pOH. The term *pH* is used to express the acidic or alkaline condition of a solution and is defined as the negative logarithm of the active hydrogen ion concentration expressed in moles per litre ($[H^+]$).

$$pH = - \log[H^+] \tag{7.1}$$

or

$$pH = \log 1/[H^+]$$

The pH scale runs from 0 to 14, with the neutral point (pH 7) being the pH of pure water at approximately 25°C (77°F). Alkaline solutions have a pH above 7, whereas acidic solutions have a pH of less than 7. Because the measure of pH is a logarithmic function, a solution having a pH value of 5 has 10 times more active hydrogen ions than a solution of pH 6. Similarly, a solution of pH 2 has 1 000 times more active hydrogen ions than a solution of pH 5.

A solution of pH 1 has 10^{-1} mole/L of free hydrogen ion concentration, whereas a solution of pH 13 has 10^{-13} mole/L of free hydrogen ion concentration. An instrumentation system that accurately measures this wide range in free hydrogen ion concentrations must be extremely sensitive at the lower range of hydrogen ion concentrations and capable of measuring a wide span of concentrations.

The hydrogen ion concentration varies inversely to the free hydroxyl ion concentration expressed in moles per litre [OH], according to the following equilibrium equation:

$$[H^+][OH^-] = 10^{-14} \tag{7.2}$$

In a litre of pure water at 25°C, approximately 10^{-7} moles of water dissociate, producing the same concentration of free hydrogen and hydroxyl ions. The negative logarithm of the hydroxyl ion concentration is designated as the *pOH*. The relationship between pH and pOH can be derived from Equation (7.2) by taking negative logarithms of both sides to obtain

$$pH + pOH = 14 \tag{7.3}$$

Acids and bases dissociate in water, producing hydrogen and hydroxyl ions, respectively. An acid is described as weak or strong depending on the number of hydrogen ions liberated when a given amount of acid is added to water. A base is also described as strong or weak depending on the number of hydroxyl ions liberated when the base is added to water. Nitric acid (HNO_3), for example, is defined as a strong acid because nearly all of the acid molecules dissociate in water to produce hydrogen ions and nitrate ions. Conversely, a weak acid like acetic acid (CH_3COOH) dissociates very little and produces few hydrogen ions in aqueous solution.

If acid is added to water, additional hydrogen ions are produced and the concentration of the hydrogen ions increases. If 10^{-3} moles of a strong acid, such as hydrochloric acid (HCl), are added to a litre of water, nearly all of the acid dissociates, producing about 10^{-3} mole of hydrogen ions and a pH of approximately 3.

If a base is added to water, hydroxyl ions are produced and the number of hydrogen ions decreases, while the pH increases. For example, if 10^{-2} moles of the strong base sodium hydroxide (NaOH) are added to a litre of water, the hydroxyl ion concentration will be 10^{-2} moles/L and the pH will be approximately 12 (that is, using Equation 7.3, $12 = 14 - 2$).

ACIDITY AND ALKALINITY. Concepts of acidity and alkalinity are useful for determining neutralization requirements. More than the measurement of pH is required to adequately determine how much base is needed to neutralize an acid or how much acid is needed to neutralize a base (Hoffman, 1972). A simple pH measurement of a nitric acid solution and an acetic acid solution will not yield identical results even though similar amounts of the acids are present in each solution. Furthermore, the pH measurement will not define how much base will have to be added to neutralize the acid in both solutions. In the nitric acid solution, almost all of the hydrogen ions that must be neutralized are present initially as hydrogen ions and are quantified by the pH measurement. However, acetic acid in solution is ionized weakly into hydrogen ions and acetate ions (CH_3COO^-). As free hydrogen ions combine with added hydroxyl ions to form undissociated water, more acetic acid dissociates to maintain the hydrogen ion concentration that exists at equilibrium in the solution. Therefore, a pH measurement alone will not yield full information about how much base must be added to neutralize the acetic acid solution.

A solution of a strong acid may have a much lower initial pH than a solution of a weak acid, even if equivalent amounts of each are used to prepare the individual solutions. The total acidity of the two solutions, however, will be the same, and equal amounts of a base will be required to neutralize equal volumes of the two solutions. Conversely, a dilute solution of a strong acid may have a pH similar to a concentrated solution of a weak acid and yet require a significantly greater dosage of a base to neutralize, even if the volumes of the two solutions are equal.

Acidity. Acidity of a wastewater is a measure of its capacity to neutralize a strong base (NaOH) to a designated pH. Acidity is expressed as an equivalent amount (in milligrams per litre) of calcium carbonate ($CaCO_3$). It is a gross measure of a wastewater property and can be interpreted in terms of specific dissolved substances only when the chemical composition of the wastewater is known. Strong mineral acids like sulfuric acid (H_2SO_4) and weak acids like carbonic (H_2CO_3) or acetic acids, as well as metal salts like ferrous or aluminum sulfate, contribute to the measured acidity of the wastewater.

The measured value of acidity varies with the designated endpoint pH used in its determination. In the titration of a single acidic species, the most accurate endpoint pH for acidity is obtained from the inflection point of a titration curve. The inflection point is the point at which the slope of the curve (pH change per millilitre of added reagent) is greatest. The point is determined by inspection of the titration curve. Because accurate identification of the inflection point of a curve is difficult in buffered or complex wastewater mixtures, the titration in such cases is carried to an arbitrary endpoint pH. The selected endpoint pH however, has been standardized as either the "methyl orange end point" or the "phenolphthalein end point" (APHA, 1992). The methyl orange endpoint pH is approximately 4.3 and the phenolphthalein endpoint pH is 8.2 to 8.4. The amount of base required (expressed as $CaCO_3$) to raise the wastewater pH from its initial value to the methyl orange endpoint is defined as the *methyl orange acidity* or the *mineral acidity*. The amount of base required to raise the wastewater pH from its initial value to the phenolphthalein endpoint is defined as the *phenolphthalein acidity* or the *total acidity*. Therefore, wastewaters with low pH (< 4.3) will have both methyl orange acidity and phenolphthalein acidity, whereas those at higher pH values ($4.3 \leq pH \leq 8.5$) will have only phenolphthalein acidity. Wastewaters with initial pH > 8.5 have no measurable, that is, titratable acidity.

In wastewaters with no mineral acidity (pH > 4.3), carbon dioxide, a normal component of all natural waters, may furnish a large portion of any titrated acidity. Carbon dioxide may be produced in waters by biological oxidation of organic matter. The carbon dioxide equilibrium in natural waters affects the amount of acidity and alkalinity measured by titration. Salts of weak bases also contribute to the measured acidity because they consume base when titrated to the designated endpoint pH values.

Alkalinity. The alkalinity of a wastewater is a measure of its capacity to neutralize a strong acid (such as H_2SO_4) to a designated pH. Alkalinity, like acidity, is expressed as an equivalent amount (in milligrams per litre) of calcium carbonate. It is a gross measure of a wastewater property and can be interpreted in terms of specific substances only when the chemical composition of the sample is known. Alkalinity of many wastewaters is primarily a function of the carbonate (CO_3^{-2}), bicarbonate (HCO_3^-), and hydroxide equilibria of the wastewater. The alkalinity is taken as an indication of the combined concentrations of these constituents and may also include contributions from borates, phosphates, silicates, and other anions.

Alkalinity is measured by titration of the wastewater with a solution of dilute sulfuric acid. Samples of an initial pH above 8.3 have the titration made in two steps. The first titration step is made to the phenolphthalein endpoint pH (8.3 to 8.5) and is designated as the *phenolphthalein alkalinity*. The second phase of the titration is conducted to the methyl orange endpoint pH (4.3) and is designated as the *methyl orange alkalinity*. The total alkalinity is the sum of the methyl orange alkalinity and phenolphthalein alkalinity. Wastewater of pH < 4.3 has no measurable alkalinity.

Normally, domestic wastewater has an alkalinity only slightly greater than that of the same domestic water supply. The alkalinity of industrial wastewater, however, can vary widely and should never be assumed to be the same as the domestic water supply.

BUFFERING CAPACITY. Buffering capacity is defined as the capacity of a solution to resist changes in pH. It results from the presence of weak acids and their salts or weak bases and their salts. Compounds such as carbonate, bicarbonate, and salts of phosphoric acid ($H_2PO_4^-$ and HPO_4^{-2}) provide a good buffer system in wastewater. Strong acid with strong base reactions, however, are more common in industrial operations and provide negligible buffering capacity to the wastewater. In some operations, the buffer capacity normally present in the raw water has been destroyed by the addition of strong acids or bases or by intentional softening of the water for process use. A wastewater with a low buffering capacity is difficult to neutralize and maintain within a relatively narrow pH range because addition of small quantities of the neutralizing agent can make large changes in the pH. Addition of a buffering agent to the waste stream may result in a more economical and controllable neutralization system.

WASTEWATER CHARACTERIZATION

Municipal sewerage agencies have sewer-use ordinances that establish allowable pH ranges for discharge to their collection systems. These limits must be met by industrial dischargers to prevent problems in the wastewater collection system or at the WWTP. The limiting discharge values vary with the individual agency but are typically pH 5 to 9 or pH 5 to 10. Some municipalities do not specify an upper pH limit because higher pH values are not as detrimental to the sewerage system as are lower pHs. The General Pretreatment Regulations (40 CFR Part 403) specify, however, that the lower pH limit cannot be less than 5.0.

TITRATION CURVES AND ANALYSIS. Though the pH value itself is a control point for the neutralization process, the acidity or alkalinity of the wastewater is the true measure of the quantity of either acid or base required for neutralization. To determine the total acidity or alkalinity of a solution, a titration must be performed as previously described. Such titration analyses are a necessary first step in the design of a pH neutralization system. The data obtained define the required neutralizing agent, the expected dosage, and the system control characteristics. An acid–base titration curve is a plot of pH versus the amount of reagent added and graphically shows the pH changes per unit addition of reagent. The shape of the titration curve depends on such variables as

- Nature of the acid or base present (for example, strong or weak);
- Concentration of the acid or base;
- Other buffers present; and
- Type and concentration of the reagent used for pH adjustment (Hoffman, 1972).

The effects of these characteristics on pH control requirements are indicated by the "shape" of the titration curve. The equivalence or inflection point is the point on the curve where the pH changes most dramatically per unit of reagent added. If a strong acid is mixed with a strong base, the equivalence point is at pH 7, but if weak acids and bases are mixed, the equivalence point is above or below pH 7. Figure 7.1 shows a representative titration curve of a strong acid, such as sulfuric, nitric, or hydrochloric acid, titrated with a strong base, such as sodium hydroxide. Near the equivalence point (point at which there are equivalent amounts of acid and base in solution), the titration curve is steep and a slight excess of caustic results in a change in pH of several units. Because the wastewater-neutralizing reagent system provides no buffering capacity, maintenance of a narrow ban for pH control of

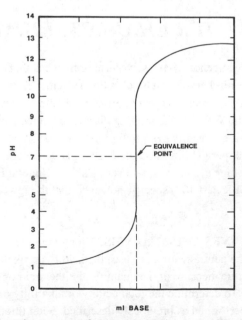

Figure 7.1   Titration curve of strong acid titrated with a strong base.

the treated effluent is difficult. Therefore, a pH control system for neutralizing a strong acid–strong base mixture must provide chemical feed and pH-sensing equipment that are designed specifically for this application.

The equivalence point of the titration of a strong acid with a weak base occurs at pH less than 7, as shown in Figure 7.2. The pH change per unit of base added is not nearly so pronounced as in the strong acid–strong base titration. Strong acids reacted with weak bases and strong bases reacted with weak acids typically produce salts that provide buffering. Because of this, the neutralization of a strong acid with a weak base or *vice versa* typically results in better pH control. In these cases, it is easier to maintain the desired effluent pH, which may not be attainable using a strong acid–strong base neutralization system. An advantage of using the salt of a weak base, such as sodium bicarbonate (NaHCO$_3$), for neutralization of a wastewater containing strong acid is that the weak base bicarbonate ion (HCO$_3^-$) provides buffer capacity and produces a flatter titration curve within the acceptable range of pH control.

A well planned and comprehensive wastewater-sampling program should be conducted to provide accurate information on the type and amount of reagent necessary to neutralize high- or low-pH solutions. Wastewater samples that are representative of the full range of expected wastewater variation should be obtained, and titration curves should be developed to determine the maximum required chemical doses for pH control.

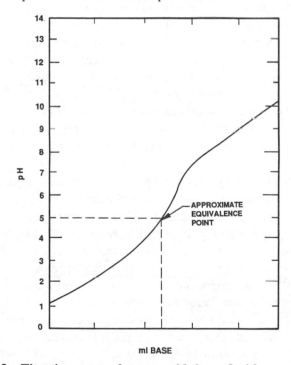

Figure 7.2 Titration curve of strong acid titrated with a weak base.

EFFECTS OF FLOW AND LOAD VARIABILITY. Adequate pH control depends on the following factors:

- Wastewater flow rate,
- Wastewater acidity or alkalinity,
- Buffering capacity of the wastewater,
- Type and strength of the neutralizing acid or base,
- Method of adding the neutralizing agent,
- Accuracy of required pH control,
- Physical system layout, and
- Variability in the pH instrumentation system.

To satisfy specific pH control requirements, each pH control system should be individually developed and designed because a change of one pH unit represents a tenfold change in hydrogen ion concentration.

The design of neutralization systems should also consider the effects of variations in wastewater flow rate, pH, and buffering capacity as expressed by the acidity or alkalinity. If these variations exist, an equalization process placed upstream of the neutralization step can facilitate pH control. Equalization is effective at dampening variations in wastewater characteristics. Mathematical models of equalization of conventional wastewater characteristics such as dissolved solids, suspended solids, or biochemical oxygen demand and chemical oxygen demand are straightforward because they can be considered conservative under normal conditions prevailing in the equalization tank. Modeling the equalization of pH, however, is complicated by the fact that pH is not a conservative substance but a result of multiple complex chemical equilibria, which may change rapidly as a result of sudden inputs of certain chemicals. Furthermore, equalization for pH control is less effective than for the other characteristics because a wastewater pH change of only two units represents a 100-fold change in the wastewater parameter. Changes of this magnitude are seldom observed for the other parameters subject to equalization.

When acidic and alkaline wastes are generated at the same or neighboring locations within the production area or plant sites, combining the waste for self-neutralization purposes can be cost effective. Storage tanks for the separate wastes may be needed to provide the proper volumes for blending and to avoid slugs of either acid or alkali. Provisions should also be made for supplementing the weaker waste stream, which may be inadequate to neutralize the stronger one. When blending waste streams from multiple industrial locations, however, overall compatibility must be evaluated by considering factors other than mutual neutralization capacities. This can be done by reviewing material safety data sheets, contacting chemical suppliers, and performing simple laboratory testing and analyses.

SLUDGE PRODUCTION. Sludge is produced from precipitation of certain compounds as a result of a change in pH. The following types of sludge are most frequently encountered in neutralization processes.

Calcium sulfate (gypsum; $CaSO_4$) is slightly soluble (2 to 3 g/L [0.07 to 0.11 oz/gal]), and is generated as a result of neutralizing alkaline wastewater containing dissolved calcium with sulfuric acid, or neutralizing acidic wastewater containing sulfates using lime compounds.

Calcium chloride ($CaCl_2$) is highly soluble and generated as a result of neutralizing alkaline wastewater containing dissolved calcium with hydrochloric acid, or neutralizing acidic wastewater containing chlorides using lime compounds. Because of its high solubility, it is likely that the available calcium (Ca) will precipitate with other salts (such as $CaSO_4$ and/or $CaCO_3$) rather than with chloride.

Metal hydroxides, including hydroxides of heavy metals, will precipitate upon neutralization of acidic wastewater containing dissolved metals using lime or caustic soda.

Some metals that normally precipitate as hydroxides at higher pH are co-precipitated with another metal hydroxide that is formed at lower pH. Ferric hydroxide, for example, coprecipitates many other metals at a pH of approximately 7.

Calcium carbonate ($CaCO_3$) deposits (scale) are generated when an over-saturation of $CaCO_3$ exists in solution and the pH values exceed approximately 8.5. The concentration of both calcium (hardness) and carbon dioxide (biological action) in the wastewater promotes the formation and precipitation of $CaCO_3$.

Sludge produced during neutralization must be processed; thus, it affects the overall cost of neutralization. The total cost of chemicals plus sludge handling and disposal using an inexpensive neutralizing agent may be greater than the total cost of using a more expensive neutralization agent which produces less sludge and has lower associated sludge-handling and disposal costs.

ALTERNATIVE NEUTRALIZING AGENTS

The criteria discussed below should be used in the selection of neutralization chemicals. Evaluation of these criteria should be based on bench-scale jar tests and preparation of titration curves under varying test conditions using representative wastewater samples.

- Type of neutralizing agent required—alkali or acid—this is the first question to answer in the selection process. Frequently, both types of neutralizing agents may be needed, especially when the pH must be controlled within a narrow range and waste stream characteristics are highly variable.

- Cost and dosage—an economic evaluation of neutralizing agents requires consideration of several alternative agents. The unit cost of each chemical and the required dose must be determined.
- Equipment cost—instrumentation (type and complexity) and the required materials of construction for tanks, mixers, and pumps are often the most significant considerations affecting the capital cost of a neutralization system.
- Residence time (reaction speed)—the time required for the neutralization reaction to be completed will affect the size of neutralization vessels.
- Production of dissolved solids—the concentration of dissolved solids produced in the neutralization process varies with the type and amount of chemicals used. Soluble salts may be objectionable in the effluent.
- Sludge—the amount of sludge generated by different neutralizing agents will vary considerably. This sludge will either be discharged to downstream processes in suspended form or will have to be periodically removed, dewatered, and disposed of. The dewatering characteristics of sludge vary with the types of neutralizing chemicals used; thus, sludge characteristics dictate the size and type of dewatering equipment. High sludge production typically results in scale formation and potential operational and process control problems.
- Safety—some neutralizing agents must be handled with greater caution than others. Precautions that must be taken to reduce skin contact, accidental eye contact, and vapor inhalation should be considered in the selection process.
- Maximum pH in overtreatment—this aspect is especially important for neutralization processes that are used before or as part of biological treatment. With some neutralizing agents, such as caustic soda, alkaline pH excursions can occur if the agent is overdosed. Overdosing with agents such as magnesium hydroxide, however, will not cause a significant pH increase because the agent becomes insoluble at pH > 9.0.
- Ease of handling—solids- or slurry-handling equipment for the neutralizing agent typically are more cumbersome and maintenance intensive than liquid-handling equipment. Use of some agents requires special equipment such as heating for bulk storage and coils for pipes, valves, and vessels. Gas and dust generation during chemical handling must also be considered.
- Availability—the local availability of some chemicals and price volatility are also important considerations.

Sulfuric acid is the most commonly used agent for neutralizing excessive alkalinity. Carbon dioxide is a weak acid and can be used for pH control if the wastewater's buffer capacity is low and proper contacting equipment is available. For acidic wastes, lime, sodium hydroxide, and sodium carbonate are used as neutralizing agents. The most commonly used neutralization chemicals are described below and a more complete list is presented in Table 7.2. Table 7.3 presents information on bulk chemical properties and handling and feeding requirements.

BASIC AGENTS. The most common bases used for neutralization of acidic wastes are discussed below.

Lime. Lime in various forms, because of its availability and relatively low cost, is the most commonly used base to neutralize acid. Its principal disadvantage is sludge and scale production from the formation of insoluble calcium salts. The resulting sludge requires removal in a clarifier or pond dewatering and subsequent disposal. Even with small concentrations of sulfate, insoluble calcium sulfate will precipitate if sufficient lime is added. Insoluble calcium salts are a maintenance concern because they necessitate frequent cleaning of pH electrodes, valves, pipes, pumps, and weirs. Lime compounds dissolve and react slowly and require relatively long contact times and high mixing power levels to promote the necessary neutralization reactions. Lime neutralization of strong acids to pH levels below 5 seldom exhibits a scaling problem. At pH levels between 5 and 9, the deposits formed may be either granular sludge or scale depending on the lime used, whether wastewater solids are recirculated as seed, and the wastewater characteristics. Lime addition taking the pH to the range of 9 to 11 may cause formation of a strongly adhering and hard scale that must be removed to prevent fouling of the system.

There are a variety of lime compounds used in neutralization of acid wastes, including

- High-calcium hydrated lime,
- Calcium oxide (unslaked lime),
- Dolomitic quicklime,
- Dolomitic hydrate,
- High-calcium limestone,
- Dolomitic limestone,
- Calcium carbonate, and
- Spent calcium carbide waste (calcium hydroxide).

Each compound has a different *basicity factor*, as indicated in Table 7.2. The basicity factor is defined as the gravimetric equivalent of a material's ability to neutralize acid. The several forms of lime also have different

Table 7.2 Neutralization factors for common alkaline and acid reagents.

Chemical	Formula	Equivalent weight	To neutralize 1 mg/L acidity or alkalinity (expressed as $CaCO_3$) requires n mg/L	Neutralization factor, assuming 100% purity of all compounds
				Basicity
Calcium carbonate	$CaCO_3$	50	1.000	1.000/0.56 = 1.786
Calcium oxide	CaO	28	0.560	0.560/0.56 = 1.000
Calcium hydroxide	$Ca(OH)_2$	37	0.740	0.740/0.56 = 1.321
Magnesium oxide	MgO	20	0.403	0.403/0.56 = 0.720
Magnesium hydroxide	$Mg(OH)_2$	29	0.583	0.583/0.56=1.041
Dolomitic quicklime	$[(CaO)_{0.6}(MgO)_{0.4}]$	24.8	0.497	0.497/0.56 = 0.888
Dolomitic hydrated lime	$\{[(Ca(OH)_2]_{0.6}[Mg(OH)_2]_{0.4}\}$	33.8	0.677	0.677/0.56 = 1.209
Sodium hydroxide	$NaOH$	40	0.799	0.799/0.56 = 1.427
Sodium carbonate	Na_2CO_3	53	1.059	1.059/0.56 = 1.891
Sodium bicarbonate	$NaHCO_3$	84	1.680	1.680/0.56 = 3.000
				Acidity
Sulfuric acid	H_2SO_4	49	0.980	0.980/0.56 = 1.750
Hydrochloric acid	HCl	36	0.720	0.720/0.56 = 1.285
Nitric acid	HNO_3	62	1.260	1.260/0.56 = 2.250
Carbonic acid	H_2CO_3	31	0.620	0.620/0.56 = 1.107

Table 7.3 Summary of properties for typical neutralization chemicals.

Property	Calcium carbonate ($CaCO_3$)	Calcium hydroxide ($CA(OH)_2$)	Calcium oxide (CaO)	Hydrochloric acid (HCl)	Sodium carbonate (Na_2CO_3)	Sodium carbonate ($NaOH$)	Sulfuric acid (H_2SO_4)
Available form	Powder, crushed (various sizes)	Powder, granules	Lump, pebble, ground	Liquid	Powder	Solid flake, ground flake, liquid	Liquid
Shipping container	Bags, barrel, bulk	Bags (50 lb),[a] bulk	Bags (80 lb), barrels, bulk	Barrels, drums, bulk	Bags (100 lb), bulk	Drum (735, 100, 450 lb)	Carboys, drums (825 lb), bulk
Bulk weight (lb/ft^3)	Power 48 to 71; crushed, 70 to 100	25 to 50	40 to 70	27.9%–0.53 lb/gal,[b] 31.45%–9.65 lb/gal	34 to 62	Varies	106, 114
Commercial strength	—	Normally 13% $CA(OH)_2$	75 to 99%, normally 90% CaO	27.9%, 31.45%, 35.2%	99.2%	98%	60° Be–77.7% 66° Be–93.2%
Water solubility, (lb/gal)	Nearly insoluble	Nearly insoluble	Nearly insoluble	Complete	0.58 at 32°F, 1.04 at 50°F, 1.79 at 68°F, 3.33 at 86°F[c]	3.5 at 32°F, 4.3 at 50°F, 9.1 at 68°F, 9.2 at 86°F	Complete
Feeding form	Dry slurry used in fixed beds	Dry or slurry	Dry or slurry (must be slaked to $CA(OH)_2$)	Liquid	Dry, Liquid	Solution	Liquid
Feeder type	Volumetric pump	Volumetric metering pump	Dry-volumetric, wet slurry (centrifugal pump)	Metering pump	Volumetric feeder, metering pump	Metering pump	Metering pump
Accessory equipment	Slurry tank	Slurry tank	Slurry tank, slacker	Dilution tank	Dissolving tank	Solution tank	—
Suitable handling materials	Iron, steel	Iron, steel, plastic, rubber hose	Iron, steel, plastic, rubber hose	Hastelloy A, selected plastic and rubber types	Iron, steel	Iron, steel	—
Comments	—	—	Provide means for cleaning slurry transfer pipes	—	Can cake	Dissolving solid forms generate much heat	Provide for spill cleanup and neutralization

[a] lb × 0.453 6 = kg.
[b] lb/gal × 0.119 8 = kg/L.
[c] 0.555 (°F − 32) = °C.

reaction times that may significantly affect the size of neutralization tanks and, therefore, capital costs. Some sources of lime may have a significant percentage of inert materials that add to the quantity and type of sludge produced.

Limestone is the least expensive of the available acid-neutralizing chemicals but it has serious limitations. When used in beds through which the waste stream passes, it reacts to produce carbon dioxide, which tends to gas-bind the beds. The coating of calcium sulfate formed on the bed material by sulfuric acid waste must be removed by mechanical agitation. The reaction times may be as much as 1 hour or longer, depending on the quality and size of the stones. Limestone has to be periodically removed to maintain process effectiveness. Overtreatment with lime can raise the pH to as high as 12.5.

Limestone has also been used successfully in rotating horizontal drums but mainly as a pretreatment process before "fine-tuning" for pH control.

Caustic Soda. Anhydrous caustic soda (NaOH) is available but its use typically is not considered practical in wastewater treatment applications because of potential safety problems associated with material handling and dissolution. Therefore, only liquid caustic soda is discussed herein.

Caustic soda is expensive but offers numerous advantages in terms of system capital and operation and maintenance costs. Caustic is a strong neutralizing agent and has a fast reaction rate, thereby reducing tankage requirements. It is also a "clean" chemical in terms of feed and storage requirements and results in significantly less sludge production than lime-based compounds. The sodium salts formed when using caustic soda usually are highly soluble so sedimentation after pH adjustment may not be required.

The most significant concern with caustic soda is that it is generally co-produced during the manufacture of chlorine. Because the supply of chlorine varies by region from surplus to shortage, the price of caustic soda is volatile and its availability varies accordingly. Environmental concerns related to the use of chlorine are likely to reduce the demand for chlorine and, therefore, reduce the supply and increase the price of sodium hydroxide. Another disadvantage is that it demands strict attention to health and safety concerns for storage and handling. Caustic fumes are harmful to unprotected skin.

Sodium Bicarbonate. As a salt of the weak acid H_2CO_3, sodium bicarbonate (NaHCO$_3$), or baking soda, is a highly effective buffering agent. Although not a strong neutralizing agent, it is quite useful for adding alkalinity (or acidity) because it provides the bicarbonate species at a pH near neutrality. Its use in neutralization is primarily as a buffering agent. It is particularly effective for pH and alkalinity control in anaerobic treatment.

Sodium Carbonate. Sodium carbonate (Na_2CO_3; soda ash) can be handled with fewer and less extreme precautions than caustic soda when used as a neutralizing agent. It is also less expensive than sodium bicarbonate but generally less effective than either NaOH or NaHCO$_3$ when used as a neutralizing agent. Because of its low solubility in water, soda ash is most economically fed as a slurry in the same manner as hydrated lime. Soda ash is a moderately fast-acting neutralization agent but generates carbon dioxide, which may cause foaming problems.

Magnesium Hydroxide. Magnesium hydroxide [Mg(OH)$_2$] is a weak base, relatively safe to handle, and unlike lime and caustic soda, is endothermic upon dissolution in water. The chemical has gained a degree of acceptance as a cost-effective alternative for neutralizing acidic streams, especially when dissolved metals must be removed. The use of Mg(OH)$_2$ generally results in low-volume metal hydroxide sludge. The sludge, however, is more difficult to dewater than sludge generated by addition of lime compounds. The solubility of magnesium hydroxide is low at ambient temperatures and decreases as the temperature increases. As a neutralizing agent, magnesium hydroxide has a high basicity but does not react as rapidly as either lime or caustic soda. It becomes insoluble at pH 9.0, thus preventing accidental pH excursions resulting from overdose. In contrast, caustic soda can rapidly increase the pH of a waste stream to greater than 12 if it is overdosed. This may present not only a process safety concern, but also a regulatory compliance concern under the Resource Conservation and Recovery Act, as pH 12.5 defines a corrosivity hazard.

ACIDIC AGENTS. The most common acids used for neutralization of alkaline wastewater are discussed below.

Sulfuric Acid. Sulfuric acid (H$_2$SO$_4$) is the most commonly used chemical for neutralizing alkaline wastewaters. Soluble sodium salts and insoluble calcium salts are produced when the alkaline waste is high in sodium and calcium, respectively. Sulfuric acid is economical and requires conventional materials for storage and feeding. However, special safety and materials-handling precautions must be taken because of its corrosiveness. In addition, the sulfate (SO$_4$) ion can be reduced to sulfide under anaerobic conditions in the sewer and result in hydrogen sulfide (H$_2$S) production.

Carbon Dioxide and Flue Gas. Compressed carbon dioxide (CO$_2$) gas for neutralization of alkaline wastewater has become more popular in recent years. The gas is a byproduct of ethanol production. When CO$_2$ is dissolved in the wastewater, it forms the weak acid H$_2$CO$_3$, which reacts with caustic wastes and neutralizes them to the target pH. Neutralization with CO$_2$ is most cost-effective when used to "fine tune" the wastewater pH in a two- (or more) stage neutralization process or when only minor pH adjustment is required.

Use of flue gas for neutralization of alkaline wastewater may be an economical method of CO_2 addition, depending on its availability. The flue gas contains approximately 14% CO_2. Neutralization with flue gas uses the same chemical principles as neutralization with compressed CO_2 gas.

Other Acids. Other acids, such as hydrochloric or nitric acid, are also used for neutralization but are more expensive and more difficult to handle than H_2SO_4.

COST PER UNIT ACIDITY/ALKALINITY. The cost of chemical to provide a unit of alkalinity or acidity is the most appropriate basis on which to evaluate alternative neutralizing agents. The following equation can be used for this evaluation:

$$\begin{array}{c}\text{Dollars per metric ton} \\ \text{of alkalinity or acidity} \\ \text{as } CaCO_3\end{array} = \frac{\text{(Cost in dollars per metric ton)(Equivalent weight)}}{\text{(Fractional purity of bulk chemical)(50)}}$$

$$(7.4)$$

BULK STORAGE REQUIREMENTS. General material and bulk storage requirements for alternative neutralization chemicals are given below. Chemical suppliers, manufacturers, and their trade associations should be consulted before selecting materials and storage facilities for these chemicals.

Bulk quicklime (CaO) is stored in airtight concrete or steel bins having at least a 60-deg slope on the bin outlet. Hopper agitation in storage bins is generally not required. Bulk lime can be conveyed by conventional bucket elevators and screw, belt, apron, drag-chain, and bulk conveyors of mild steel construction. Pneumatic conveyors subject lime to air slaking, and particle size may be reduced by attrition. Dust collectors should be provided on manually and pneumatically filled bins. Undissolved grit in the hydrated lime slurry can cause excessive wear on valves, pumps, and other handling and storage equipment.

Storage requirements for hydrated lime [Ca(OH)$_2$] are the same as for quicklime except that hopper agitation of the storage bin outlet should be provided. Also, bin outlets should be provided with nonflooding rotary feeders and hopper slopes should be 65 deg minimum. Undissolved grit in the hydrated lime slurry is less than for quicklime but can cause excessive wear on valves, pumps, and other handling and storage equipment.

Liquid caustic soda (NaOH) can be stored at a 50% concentration (by weight). At this solution strength, however, it crystallizes at 11.7°C (53°F). Therefore, storage tanks must be located indoors, or, if outdoors, provided with heating and suitable insulation. Alternatively, liquid caustic soda can be diluted to 20% solution (by weight), which crystallizes at 28.9°C (84°F). Recommendations for dilution of both 73% and 50% solutions should be obtained from the manufacturer because special handling and safety considerations are necessary.

Storage tanks for liquid caustic soda should be provided with an air vent to allow gravity flow. The storage capacity typically is approximately 1.5 times the largest expected bulk delivery, with an allowance for dilution water if used or 2 weeks supply at the anticipated feed rate, whichever is greater. Tanks for storing 50% solution at temperatures between 24°C (75°F) and 60°C (140°F) may be constructed of mild steel. Storage temperatures higher than 60°C (140°F) require more elaborate materials selections and are not recommended. Caustic soda will tend to pick up iron when stored in steel vessels for extended periods. Subject to temperature and solution limitations, rubber, 316 stainless steel, nickel alloys, or plastics may be used when iron contamination must be avoided.

Soda ash (Na_2CO_3) usually is stored in steel bins and conveyed by steel pneumatic equipment provided with dust collectors. Bulk and bagged soda ash tend to absorb atmospheric carbon dioxide and water, forming the less active sodium bicarbonate. When soda ash is stored as a slurry, it is often convenient to pump it directly from the unloading point to the storage tank. Slurries containing 50 to 60% total soda ash (by weight) can be pumped but require special care (preventing heat loss) to avoid crystallization. Weak solutions containing 5 to 6% dissolved soda ash can be handled the same as water. Arching, bridging, or "rat-holing" is sometimes encountered when storing light soda ash. This can be overcome by installing electric or pneumatic vibrators mounted on the outside of the bin bottom just above the outlet. The use of vibrators is not recommended with dense soda ash.

A slurry storage system consists of a tank, a means of slurrying the bulk soda ash and transferring it to storage, and a means for reclaiming solution from the tank and replenishing the water. One of the most important requirements for successful operation of a slurry storage system is maintaining the required operating temperature in the slurry bed to prevent formation of crystals, which are difficult to redissolve. Water used for operating the system is preferably preheated. Heating coils may be immersed in the bottom of the slurry tank. If the slurry tank is located outdoors, insulating may be necessary.

Sodium bicarbonate ($NaHCO_3$) storage requirements in dry form are similar to those for soda ash (see above).

Magnesium hydroxide [$Mg(OH)_2$] is available as an aqueous slurry of agglomerated particles at 55 to 60% $Mg(OH)_2$. This slurry freezes at 0°C (32°F) and must be kept in mildly agitated storage. While the product should not be harmed by freezing, separation may occur and reslurrying may be difficult.

Baume sulfuric acid 66° (H_2SO_4) does not attack iron and, therefore, can be stored in unlined steel tanks containing an air vent fitted with a dryer. If the concentration of the acid is less than 66° Baume, it is advisable to line the tank with an acid-resistant coating.

Liquid system capacities encountered in wastewater treatment usually require on-site bulk storage units. Standard prepackaged units are available, ranging in size from 0.25 to 51 metric tons (0.25 to 50 tons) capacity, and are furnished with temperature–pressure controls to maintain approximately 2 068 500 Pa (300 lb/sq in) at −17.8°C (0°F) conditions. The typical package unit contains refrigeration, vaporization, safety, and control equipment. The units are well insulated and protected for outdoor use. The gas from the evaporation unit usually passes through two stages of pressure reduction before entering the gas feeder to prevent the formation of dry ice.

DESIGN OF pH-CONTROL SYSTEMS

A pH-control system measures the pH of the solution using a sensor element and controls the addition rate of a neutralizing agent on demand to maintain the effluent within certain acceptable pH limits. The pH-control system performs a continuous titration, except that the strength of the solution being neutralized is of no interest.

The design of control systems is complicated because pH is a nonlinear function of concentration. For example, adding a certain amount of base to a solution of strong acid of pH 2 will increase pH to 3. If it is desirable to take this solution to pH 4, only about 10% of the original base amount may be required. If it is necessary to take the pH on to 5, only about one 1% of the first amount of base is added; and for pH 6, only about 0.1% of the first amount is required. Taking a waste stream from pH 2 to pH 7 is a difficult control problem. A large quantity of base must be added before any change in the pH is produced, but as a small additional amount is added, the pH rises rapidly, depending on the buffering capacity of the wastewater. To obtain precise pH control under such conditions, an accurate and responsive control system is required.

With the introduction of microprocessor-based technology, many industrial instrumentation controllers and analyzers now use configurable algorithms to characterize their function curves. These algorithms are programmed to generate a segmented characterized curve that is inversely proportional to the titration curve for the wastewater and the selected neutralizing agent. This results in a near-linear controller output with respect to reagent demand. A single characterized curve will typically be sufficient for pH control, using a constant strength of neutralizing agent if the pH value varies between 4 and 10. If the control system detects a pH outside of this operating band, an alternate characterized curve can be developed and electronically switched to replace the normal curve until the pH is returned to the routine control region. For routine control of influent pH variations beyond 4 and 10, consideration should be given to using a split-ranged controller to control two chemical addition rates by using different-sized control valves. A large valve is used for pH excursions outside of the 4 to 10 pH range, and a smaller valve is used for control within the 4 to 10 pH range. Because of the nature of split-range control, while the large valve is throttling down the chemical feed rate, the smaller valve will be fully open, allowing a smooth control of chemical addition at the pH 4 and 10 boundary points.

An alternate method for control of pH is the use of an adjustable gain controller. As the measured pH deviates further from the desired pH setpoint, the controller gain increases, causing the controller output to add more chemical in proportion to the gain curve. As with the segmented characterized curve method, when the pH value deviates outside of the nominal 4 and 10 range, split-range controller output should be used.

A pH-control system may be implemented either by means of a batch system or a continuous-flow system. For the following example, only acidic waste streams typical of the majority of industrial applications will be considered in the descriptions.

BATCH-CONTROL SYSTEMS

Plants with high wastewater volumes and flow rates typically use continuous-flow proportional or multimode control systems. Plants with intermittent and/or low volumes of spent acids or bases may use batch neutralization systems or continuous, two-position (on–off) control systems. In terms of equipment requirements, the batch systems or on–off control systems are more economical for continuous pH neutralization processes than the multimode control systems. Batch neutralizations are customarily limited to situations where waste stream flow is irregular and the concentrations of the spent acid or base liquors are high and variable. Such a situation occurs when strong sulfuric or hydrochloric acid pickling liquors are dumped infrequently. When the acid liquors are spent, they are pumped into a tank that neutralizes a batch before discharge to the sewer.

Figure 7.3 shows a typical batch neutralization system. For an acidic wastewater, the batch cycle is started by opening the waste valve. When the level in the neutralization basin reaches a low level, as measured by the level controller (LC), the mixer starts and the reagent addition system receives a signal to begin operation. The pH is measured and the analog value is recorded on a chart recorder (AR). Discrete switches (AC) associated with the pH analyzer (AIT) are used to control the reagent valve. If the influent waste is below a low-low set value, the reagent valve will open to add chemicals to raise the pH. If the pH value of the waste exceeds the low value, the reagent valve will be controlled via a cycle timer (KY-"A"). This timer enables the valve to open for a specific time period and then close to allow for reaction time. The system will continue using either feed-control method until the pH is within the desired range. While the system is continuously controlling the

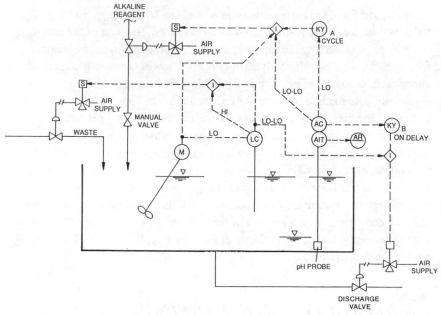

Figure 7.3   pH control schematic for batch neutralization system for acid wastewater.

waste pH, the waste valve is open, filling the basin. When the basin level reaches the high value, the waste valve will close. Adjustment of pH will still be required for a period after the waste valve is closed.

After the pH has been adjusted to within the desired limits, an on-delay timer (KY-"B") is used to delay the opening of the discharge valve. This helps ensure that the pH does not fall outside the pH limits. Once the on-delay time has timed out and the equalization basin pH is still within the limits, the discharge valve will open, allowing the basin to empty. When the basin level falls below the low value measured by the level controller (LC), the mixer will stop. The level will continue to decrease until the low-low level is reached. At this time, the discharge valve closes and the waste valve opens, beginning a new batch cycle. Note that the low-low level in the basin should be above the pH probe element to ensure that the element remains submerged.

In the system shown, the neutralizing agent typically is added through a solenoid valve or air-actuated valve. The tank typically is mixed by a propeller mixer or by the addition of air.

Segmented characterized curves can be used with batch control methods. With a known volume of wastewater in the equalization tank and measured pH, the curve can be used to set the time and, therefore, the amount of chemical addition. To enable reaction of the chemicals, this time period may be lengthened to allow a lower chemical feed rate per unit time. Neutralization of large batches still requires treatment in stages to prevent overcorrection of the pH.

CONTINUOUS FLOW SYSTEMS

The two-position, or "on–off," system (Figure 7.4) is so named because the element controlling the reagent addition is always in either the fully open or fully closed position. Wastewater continuously enters the retention basin and overflows the discharge weir. The pH value of the basin is measured using a sidestream sampling system. Discrete switches (AC) associated with the pH analyzer (AIT) are used to control the reagent valve. As long as the wastewater pH value measured in the sidestream sample piping is low, the reagent valve will be open. The analog value is recorded on a chart recorder (AR). This system is generally limited to processes where the wastewater flow rate is relatively small and the hydraulic residence or liquid holdup time within the control system is relatively large. The detention time in the reaction vessel should be a minimum of 10 minutes. Adequate mixing and agitation prevent the pH electrodes from detecting an incorrect pH, which would result in the discharge of material outside the desired pH limits. Agitation turnover time of the tank contents should be less than 20% of the hydraulic residence time. For example, if the residence time is 10 minutes, the turnover time should be less than 2 minutes. If the flow rate or the total acidity or basicity

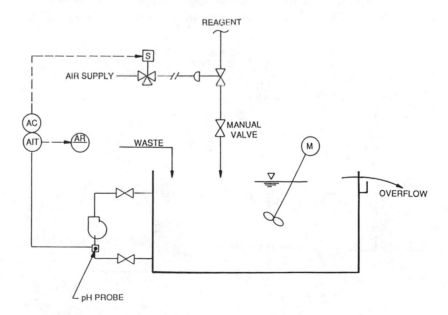

Figure 7.4 Simple on–off control switch.

of the waste stream varies by a factor of 10 000 (for example, a change of four pH units at a constant flow or a change of three pH units accompanied by a tenfold change in flow rate), then two reagent valves are needed. A large valve may be required for the gross reagent addition and a small valve for the trim addition of the reagent.

When the volume of spent acid or base is relatively high, it is impractical to provide the long hydraulic detention time required by on–off control systems. Systems for these conditions are designed with multimode control (Figure 7.5), which permits continuous flow-through of the neutralized material. Wastewater continuously enters the retention basin and overflows the discharge weir. The pH value of the waste in the retention basin is measured external to the basin in a sidestream sampling system, and the analog value is recorded on a chart recorder (AR). An analog controller (AIC) is used to control the reagent valve. The input to the controller is the measured pH value of the waste in the retention basin. The amount of neutralizing agent added depends on the deviation of the liquid's pH from an internal reference pH, that is, the "setpoint" on the pH control system. The output will throttle the pneumatically controlled reagent valve via an electronic-to-pneumatic converter (FY-I/P).

Several considerations enter into the design of such a control system. The buffering capacity of the system is the ability to absorb the neutralizing agent without a change in pH. Generally, a high buffering capacity is favorable for effective control because it levels out abrupt changes, allows adequate time for mixing, and, thus, reduces extreme changes in the position of the final control element. Unfortunately, pH neutralization systems are seldom high in

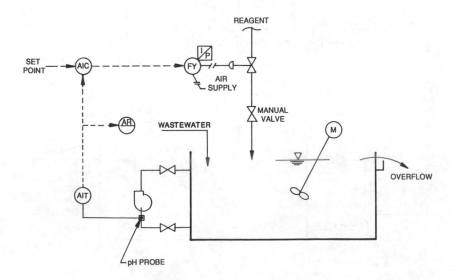

Figure 7.5   Simple multimode control system.

buffering capacity. Because pH is a logarithmic function of concentration, it is frequently desirable to use sodium carbonate or other weak bases as neutralizing agents for mineral acids. These agents offer some buffering capacity, which reduces sharp changes in pH.

Sufficient hydraulic detention time in the system is required for completion of the neutralization reaction. This is extremely important when a slurry or dry chemical feed is used as the control agent. Proper mixing is also required to eliminate delays in the development of the desired pH level. Dead volume and short-circuiting in the reaction vessel create inefficient neutralization and pH control problems.

Proper pH control can sometimes be obtained using a cascade control system with feed-forward and feed-back controls (Figure 7.6). This system should only be considered, however, when the pH variation of the influent wastewater is small. With the cascade system, wastewater continuously enters the first retention basin and overflows the discharge weir into the second retention basin. The pH value of the wastewater in the first retention basin is measured externally in a sidestream sampling system, and the analog value is recorded on a chart recorder (AR-"A"). An analog controller (AIC) is used to control the reagent valve. The input to the controller is the measured pH value of the wastewater in the first retention basin. The pH in the second retention basin is also measured using a sidestream sampling system, and the analog value is recorded (AR-"B"). The desired pH value of the waste in the first retention basin is entered as one input to a signal summer (FY-"B"). The

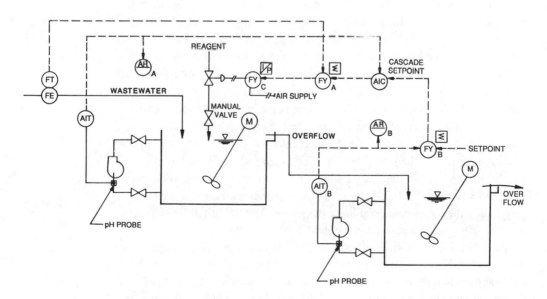

Figure 7.6   Cascade control system with feed-back and feed-forward control loops.

second input to the summer is the actual pH value in the second retention basin. The output of this summer (FY-"B") is the cascade setpoint for the analog pH controller (AIC). The output of the analog controller will be one input to another signal summer (FY-"A"). The second input to this integrator will be the feed-forward signal from the influent waste flow meter (FE/FT). The output of the other summer (FY-"A") will throttle the pneumatically controlled reagent valve via an electronic-to-pneumatic converter.

The function of the first signal summer (FY-"B") is to adjust the controller cascade setpoint up or down depending on the actual pH in the second retention basin. If the desired pH is 7.0 and the actual value in the basin is 6.5, the output of summer FY-"B" must increase, thereby increasing the output of the analog controller (AIC) and causing the reagent valve to add more neutralizing chemical to raise the pH of the first retention basin. Likewise, if the actual pH were greater than the desired value of 7.0, the signal summer (FY-"B") output would decrease the setpoint to the controller, thereby reducing the amount of reagent added to the first basin.

The function of the second signal summer (FY-"A") is to anticipate needed changes in reagent addition based on the actual amount of wastewater entering the first retention basin. If the flow rate into the basin increases, more reagent would be required for neutralization. This feed-forward signal allows the reagent valve to control changes in pH before the controller sees the addition error.

The reaction vessel dimensions should be approximately cubic, with the volume dependent on the rate of the reaction and flow rate. The inlet and outlet should be located at opposite sides of the reaction vessel to reduce short-circuiting. The reagent may be added at the same point as the influent, or it may be added to the influent before the stream reaches the neutralization vessel. An agitator should be provided to ensure good mixing, and baffles should be used in the tank to avoid a whirlpool effect. A propeller or axial flow impeller mixer or air injection may be used for mixing. The agitation should be vigorous enough that the dead time of this system is no more than 5% of the vessel's hydraulic retention time. Dead time is the interval of time between the addition of the pH adjustment reagent and the first detectable change in the pH of the effluent. A short dead time is required so that the control system's adjustment of the feed rate of the neutralization reagent is based on current information. A hydraulic detention time of at least 5 minutes should be provided with liquid neutralizing reagents. Solid neutralizing reagents require a hydraulic detention time of at least 10 minutes. If dolomitic lime is used, as much as 30 minutes detention is required.

The accuracy of the final pH adjustment depends on the range of pH control required. If the pH of the influent is 1.0 and the required pH of the effluent is approximately 7, the control system must be accurate to 1 part per million. If large changes in flow rate also occur, the control system must be accurate from 1 part per million to perhaps 1 part per 100 million. No single control valve or control element can reliably provide this degree of accuracy. An accuracy of 1 to 2% is more typical. The solution is to use upstream equalization and more than one control element, such as in the multielement system shown schematically in Figure 7.7.

For a wide pH adjustment range, three subsystems are sometimes used in series. For example, in the first stage, the pH is raised from pH 1 to approximately pH 3, a procedure that requires an accuracy of only about 1 part per 100. In the second stage, the pH would be raised to approximately 5, and in the third stage, the pH could be raised to approximately 7. Each stage of neutralization is controlled in the same manner. The actual pH value of each stage is measured and the analog value recorded on a chart recorder (AR). This measured value is the input to the analog pH controller (AIC) for each stage. The desired pH is selected at the second stage pH controller, the output of which positions the reagent valve for continuous chemical addition.

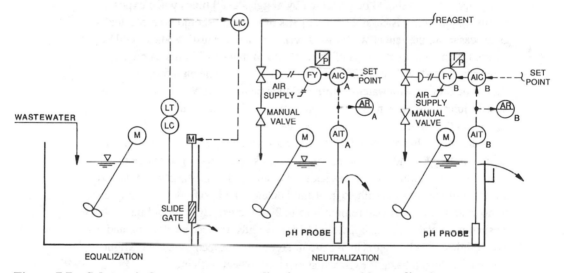

Figure 7.7   Schematic for two-stage neutralization system with equalization.

OPERATIONAL CONSIDERATIONS

As with all chemicals, manufacturers' guidelines and instructions and material safety data sheets should be consulted before use or handling. The following sections provide general guidelines for handling and use of neutralizing agents. These guidelines are not intended to be all inclusive.

RAW MATERIALS HANDLING AND SAFETY. Protective clothing and goggles are required to protect skin and eyes from CaO (quicklime). Lime dust and hot slurry can cause severe burns. Skin contacted with lime should be washed with water immediately. Lime should not be mixed with chemicals containing waters of hydration because it will react with such water, causing excessive temperatures and possible explosion. Conveyors and bins used for more than one chemical should be thoroughly cleaned before switching chemicals. Pebble quicklime is normally specified because of easier handling and less dust.

Hydrated lime [$Ca(OH)_2$] and milk of lime will irritate the eyes, nose, and respiratory system and dry the skin. Affected areas should be washed with water.

Caustic soda is a highly reactive chemical and demands great care in handling and storage. However, if handled properly, caustic soda poses no particular industrial hazard. Always add caustic soda slowly to water, and never add water to caustic. Finely diffused caustic soda fumes are very aggressive to unprotected skin. Caustic soda can cause permanent eye damage on contact; thus, the eyes should be protected by goggles at all times when exposure to mist or splashing is possible. Other parts of the body should be protected as necessary to prevent alkali burns. Areas exposed to caustic soda should be washed with copious amounts of water for 15 minutes to 2 hours. A physician should be called when exposure is severe. Caustic soda taken internally should be diluted with water or milk and then neutralized with dilute vinegar or fruit juice. Vomiting may occur spontaneously but should not be induced except on the advice of a physician.

Sodium bicarbonate ($NaHCO_3$) may cause mild respiratory irritation. It is an irritant to eyes and skin and may cause burns upon prolonged contact. Contaminated clothing and shoes should be removed immediately. Contacted areas should be washed with soap or mild detergent and rinsed thoroughly until no evidence of chemical remains (15 to 20 minutes). Eyes should be washed with water, lifting upper and lower eye lids. Protective clothing and eye goggles must be used. Gloves are not required but recommended. Slight fire and explosion hazards exist when exposed to heat or flame.

Soda ash (Na_2CO_3) is not particularly corrosive in itself, but in the presence of water, caustic soda is formed, which is quite corrosive. The dust and solution are irritating to the eyes, nose, lungs, and skin. Therefore, general precautions should be observed and the affected areas should be washed promptly with water. Dense soda ash typically is used because of superior handling characteristics. It has little dust, good flow characteristics, and will not arch the bin or flood the feeder. However, it is relatively hard to dissolve, and ample capacity must be provided to dissolve the soda ash.

Magnesium hydroxide [$Mg(OH)_2$] is comparable to a concentrated Milk of Magnesia© and, therefore, can be classified as a low-degree health hazard from a handling point of view. Contact to the eyes may cause temporary injury to the cornea. Contact with skin rarely causes irritation.

Sulfuric acid (H_2SO_4), when diluted with water, generates a considerable amount of heat; appropriate precautions must be taken. In particular, the concentrated acid must always be added to water; never is water added to the concentrated acid.

Dry carbon dioxide (CO_2) is not chemically active at normal temperatures and is a nontoxic, safe chemical. The gas however, displaces oxygen, and adequate ventilation of closed areas should be provided.

PROCESS CONTROL. Certain aspects of neutralization process control are particularly important and require ongoing evaluation and attention after the process is in place and operating.

Scale Formation. Scale deposits such as calcium carbonate and calcium sulfate form because of oversaturation conditions caused by increased pH or calcium or sulfate concentration. Scale may create problems in pipes, valves, mixers, pumps, and instrumentation sensors. This material must be periodically cleaned by manual or mechanical means or chemically dissolved. Scale that does not adhere to surfaces will settle as sludge and must be removed.

Annual Cost. The annual costs of neutralization are primarily chemical costs and equipment maintenance costs. Savings should be realized through proper chemical dosages. Titration curves should be developed regularly to ensure economic evaluation of neutralizing chemical dosages and/or operating pH control ranges. Because of the cost volatility of some neutralizing agents, this aspect of process control is particularly important. Chemical dosage should be reduced to the level required to provide reliable discharge permit compliance. Where sludges are generated, disposal costs must be factored into decision-making.

Side Reactions and Byproducts. In the event of production changes, the neutralization operation should continually be monitored for:

- Production of insoluble salts (precipitates) that must be removed;
- Soluble salts that may exceed concentrations allowed by discharge ordinances;
- Introduction to sewers of toxic metals, nitrogen, or phosphorus at unacceptable levels; and
- Production of gases in the sewer system.

Equipment and Structural Effect. Despite potentially high operating costs, neutralization must be accomplished to achieve permitted pH ranges and avoid damage to treatment plant structures, collection systems, and downstream facilities and treatment processes.

REFERENCES

American Public Health Association (1992) *Standard Methods for the Examination of Water and Wastewater.* 18th Ed., Washington, D.C.

Baker, R.J. (1974) *An Evaluation of Acid Waste Treatment.* Paper presented at Int. Water Pollut. Conf. and Expo., Detroit, Mich.

Hoffman, F. (1972) How to Select a pH Control System for Neutralizing Waste Acids. *Chem. Eng.*, **24**, 105.

Nemerow, N.L. (1971) *Liquid Waste of Industry-Theory, Practices, and Treatment.* Addison-Wesley Publishing Co., Reading, Mass.

Okey, R.W., and Chen, K.Y. (1978) Neutralization of Acid Wastes by Enhanced Buffer. *J. Water Pollut. Control Fed.*, **50**,1841.

Chapter 8
Heavy Metals Removal

Heavy metals are present in abundance naturally and enter the water cycle through a variety of geochemical processes. Many metals also are added to water by human-induced activities such as manufacturing, construction, agriculture, and transportation. At excessive concentrations, soluble metal compounds may be deleterious to health and subsequent water use. More specifically, high concentrations of heavy metals in water supplies are undesirable because of the potential adverse effects on the health of organisms, suitability of water for various purposes, longevity of water and sewer networks, and aesthetics of the environment. Pretreatment for heavy metals removal reduces potential for such adverse effects associated with passthrough or treatment inhibition. Tables 1.1 to 1.6 provide lists of U.S. Environmental

Protection Agency (U.S. EPA) toxic pollutants, conventional pollutants, and nonconventional pollutants. Metals are given specifically in Table 1.2 (refer to Sec. 307(q)(1) of the Clean Water Act).

Certain metals in low concentrations are not only harmless, but traces are essential to good nutrition (for example, cobalt, copper, iron, selenium, and zinc). Some metal salts, on the other hand, may be toxic.

Evaluations of toxicity include acute, chronic, synergistic, and mutagenic/teratogenic. Acute toxic effects show up quickly upon ingestion of, or contact with, a metal compound at dosages sufficient to cause the acute effect. Soluble copper causes symptoms of gastroenteritis with nausea. Effects produced by chromium include lung tumors, skin sensitizations, and inflammation of the kidneys. Selenium in high concentrations is a poison, a carcinogen, and a cause of tooth decay (U.S. EPA, 1979). Chronic effects develop as ingestion or contact with lower doses occur over a period of time. Some metals, such as cadmium and lead, accumulate in body tissues and are not excreted; eventually, chronic poisoning may result.

Certain metals are more toxic in combination with other metals or under specific environmental conditions. This effect is described as synergistic. For example, cadmium toxicity increases in the presence of copper or zinc. Copper, cadmium, nickel, lead, and zinc may be more or less toxic to aquatic life depending on other water quality conditions, such as pH, temperature, hardness, turbidity, and carbon dioxide content. For aquatic life, lead is more toxic if the dissolved oxygen concentration is low. Mutagenic and teratogenic toxicity may result when certain metals combine with organic compounds; these substances may produce changes in genetic makeup (mutagenicity) or abnormal tissue development in embryos (teratogenicity). At certain dosage levels of some heavy metals, carcinogenic effects have been observed. The presence of excessive heavy metals in a water supply thus may eliminate its usefulness for some of a community's water needs.

Taste, staining, and corrosion characteristics also are important considerations in the selection of a source as a drinking water supply. For example, copper in concentrations greater than 1 mg/L imparts an undesirable taste to drinking water. The same is true for iron greater than 9 mg/L and zinc greater than 5 mg/L. Iron and manganese may stain fixtures, discolor laundry, and obstruct pipes with bacteria (iron or encrustation [manganese]). Metals may interfere with industrial processes; for example, copper may cause adverse color reactions in the food industry or pit aluminum and galvanized steel. In these circumstances, the water supply must be treated before use. Certain metals, if present in irrigation water, may damage crops. The damage may be evidenced as stunted growth, plant death, or metal accumulation that renders the usable part of the plant inedible.

Because heavy metals can adversely affect water supplies, the regulatory agencies have limited the heavy metals discharged to surface streams from industrial and municipal sources. When it was understood that such metals also were harmful to the biological degradation process used at most municipal wastewater treatment plants (WWTP), the regulatory agencies began limiting the discharge of those same substances to the municipal sewer system as well. As a result, industries previously discharging excess heavy metals to the sewer now must pretreat their wastewaters.

EFFECTS ON WASTEWATER TREATMENT PLANTS

The major operational effect of heavy metals on municipal plants is decreased biological system activity caused by increased concentration of toxic heavy metals, primarily in the dissolved state. These effects are lessened with continuous exposure of biological systems to such metals. In effect, the biological system can become acclimated to the presence of what would normally be inhibitory, or even toxic, concentrations of heavy metals. On the other hand, the harmful effect may be exacerbated by changes in pH in the biological reactor, which could cause changes in the concentrations of the potentially toxic substances in the dissolved state and in the quantities that are physically or chemically adsorbed into the organisms. Thus, a shift in pH from 8 to 7 may increase the solubility of most toxic metals if they are present as hydroxides or soluble oxides, or if they are merely adsorbed to solid particles.

Whatever the exact cause, it has been proven that the presence of these heavy metals, in sufficient concentration in the influent to a biological treatment system, can decrease operating efficiency. Additionally, heavy metals tend to accumulate in the system's biological treatment operations, especially where wastewater solids are recycled. As the biologically degradable materials are acted on and converted to products of reaction (that is, carbon dioxide and water in aerobic systems, and methane and hydrogen sulfide in anaerobic systems), the metallic substances remain in the system and, thus, tend to concentrate. As a result, what was a low concentration in the influent may, in the solids produced, be converted to a concentration as high as 20 to 30 times greater.

There are additional effects that result from toxic heavy metals accumulation in the solids produced. Perhaps the most beneficial and least environmentally harmful disposal method for solids produced from biological treatment systems is land application. Although the solids produced are not high enough in nutrients to be considered fertilizers, they do have some nutrients and trace minerals, as well as substantial amounts of organic materials. If toxic heavy metals are present at elevated levels, both the annual rate of application and the total amount that may be applied to agricultural land must be controlled carefully. Otherwise, the crops produced may be unfit for use, or the production capacity of the soil may be decreased substantially.

If heavy metals pass through the WWTP in whole or in part, the WWTP may not meet its own National Pollutant Discharge Elimination System permit limitations. Government standards for industrial discharges to municipal systems place relatively uniform metals-discharge requirements on all industrial sources, whether they discharge to the surface water directly or to a municipal treatment system. Consequently, industrial dischargers of toxic heavy metals must pretreat their wastewater to control the metals before discharging to the WWTP.

As part of its comprehensive pretreatment program, a municipality also must limit heavy metals discharged to the plant. Furthermore, industrial dischargers to WWTPs must comply with additional state or federal pretreatment regulations appropriate to their industrial categories as they are proposed and implemented (for example, 40 CFR Part 403 and categorical standards).

*A*PPLICABLE INDUSTRIES

Table 8.1 lists industries and metals sometimes found in their wastewaters. Though not exhaustive, the list reflects the possible occurrences of certain typical metals in specific industries.

WATER POLLUTION CONTROL TECHNOLOGIES. The costs of complying with pollution control legislation and the increasing costs of raw material, water, and wastewater treatment have driven all industries with heavy metal contamination potential to seek ways to reduce their operating costs. Experience has shown that the two cost factors most responsive to improvements are related to the volume of water used and discharged and the amount of heavy metal that ends up in the wastewater treatment system. For example, when plating chemicals are conserved in the process stream, fewer chemicals must be replaced in the plating bath, removed from the wastewater before discharge, and disposed of in residues. Similar benefits are realized when the volume of process water is reduced: less water needs to be replaced or treated, and when wastewater is finally discharged to a public sewer system, the cost of treatment is reduced.

Heavy metal dischargers have a number of options for cost-effectively complying with water pollution control regulations. Within the three general options—modifying the process, pretreating the wastewater, and becoming a direct discharger—is a wide variety of possible modifications and treatment technologies from which to choose.

Table 8.1 Industries with possible occurrence of certain metals in industrial wastewater.

	As	Ba	B	Cd	Cr	Co	Cu	Fe	Pb	Mn	Hg	Ni	Se	Ag	Zn
Metallurgical industry	X	X	X	X	X		X	X	X	X		X	X	X	X
Nonferrous smelting	X			X	X		X	X	X	X	X	X	X		X
Metal mining	X		X	X			X	X	X	X		X		X	X
Wire drawing			X				X	X							
Metal plating			X	X	X	X	X	X	X	X		X			
Alloying	X			X		X	X			X	X	X		X	X
Foundries		X					X	X		X		X		X	X
Automotive industry					X		X	X	X						X
Glassware industry	X	X	X	X	X	X			X	X			X		
Ceramic industry	X	X	X	X	X				X	X			X		
Porcelain industry	X		X	X			X	X		X		X	X	X	X
Plastics industry				X							X				
Pesticide manufacture	X										X		X		
Organic chemicals industry	X			X				X							
Inorganic chemicals industry	X			X	X		X	X	X			X			X
Herbicide manufacture	X		X												
Fertilizer manufacture	X		X	X			X	X		X					
Detergent manufacture	X	X													
Disinfectant production				X							X			X	
Fungicide production			X								X				
Petroleum refining	X						X	X	X						
Dye manufacture	X	X		X	X		X			X		X	X		
Wood preservatives manufacture	X		X		X		X				X	X			
Hide preservatives manufacture	X										X				
Pigment formulation	X			X	X	X	X		X				X		X

Table 8.1 Industries with possible occurrence of certain metals in industrial wastewater (continued).

	As	Ba	B	Cd	Cr	Co	Cu	Fe	Pb	Mn	Hg	Ni	Se	Ag	Zn
Paint manufacture		X		X	X		X		X	X	X	X	X		X
Cosmetics/pharmaceuticals manufacture			X								X				X
Ink manufacture					X					X				X	
Animal-glue manufacture					X			X							
Tannery operations	X		X		X										
Carpet production			X	X									X		
Photographic supplies			X	X	X								X	X	
Textile manufacture				X			X	X	X		X		X		X
Pulp/paper/paperboard manufacture			X				X				X		X		X
Food/beverage processing								X						X	
Printing industry				X					X			X			X
Match production									X	X					
Battery manufacture	X			X					X	X		X			
Television tube manufacture				X					X						X
Jewelry manufacture							X							X	
Electrical/electronics manufacture		X					X		X		X		X		
Explosives manufacture		X									X				

The selection and design of a control system entails the following tasks:

- Performing a field investigation to define current achievable waste stream parameters (flow rate, pollutant types and concentrations, wastewater variability);
- Developing conceptual models of proposed treatment processes based on data obtained in the field investigation;
- Conducting bench-scale treatability studies on wastewater samples to simulate proposed treatment processes;
- Evaluating the results of these studies to assess the ability of the proposed system to meet discharge requirements;
- Using the results of this assessment to develop optimal design parameters for the full-size system; and
- Estimating capital and operating costs of the proposed system based on the operating parameters needed for adequate pollutant reduction and vendor price quotations for the equipment specified.

Choosing among the different options also requires a careful cost and benefit analysis considering

- Reliability of the systems to consistently reduce the pollutants to the levels specified in the discharge permit or other regulations;
- Investment cost of the systems;
- Operating costs of the systems including labor, utilities, wastewater treatment chemicals, and sludge disposal; and
- Savings in raw materials and sewer charges as a result of reduced water flow and reuse or recovery of metals.

Process Modifications. Process modifications are designed either to reduce water use (for example, by installing flow restrictors) or conserve chemicals (for example, by recycling or recovering the chemicals).

Some of the ways to reduce water use include

- Implementing a rigorous inspection program to discover and quickly repair water leaks;
- Installing antisiphon devices, equipped with self-closing valves, on water inlet lines;
- Using multiple counterflow rinse tanks to substantially reduce rinse water volume;
- Using spray rinses to reduce rinse water volume;
- Using conductivity cells or flow restrictors to prevent unnecessary dilution in the rinse tanks;
- Reusing contaminated rinse water and treated wastewater when feasible; and

- Using dry cleanup when possible instead of flooding with water.

Modifications to reduce the loss of chemicals, consequently reducing pollutant loads, include

- Implementing a rigorous inspection and maintenance program;
- Using spray rinses or air knives to reduce drag-out from plating baths;
- Using air agitation or workpiece agitation to improve plating efficiency;
- Recycling rinse water;
- Using spent process solutions as wastewater treatment reagents (for example, using spent acid and alkaline cleaning baths as reagents in a neutralization tank);
- Using minimum concentrations of chemicals;
- Using plating bath purification to control the level of impurities and prolong the service life of the bath; and
- Installing recovery systems to reclaim chemicals from rinse waters.

A closed-loop recovery system may be appropriate for wastes that are difficult or expensive to treat. In the case of rinse streams requiring treatment other than neutralization and clarification (containing cyanide or chromium, for example) or containing pollutants that are not effectively removed by conventional treatment (such as certain chelated metal complexes), installing a closed-loop system to recycle the rinse may reduce the water pollution control costs. A small-volume purge stream from the closed-loop recovery system will require treatment, but this should not be a major expense.

Recovery systems have been evaluated under U.S. EPA research projects (see U.S. EPA, 1985). Discussions have included the following topics:

- General,
- Evaporation,
- Reverse osmosis,
- Electrodialysis,
- Electrolytic recovery, and
- Ion exchange.

Conducting a cost analysis of each of the options is important in selecting the most appropriate process modifications for an operation. For such an analysis, the capital and operating costs of the modifications are weighed against the total benefits from reductions in raw material losses, wastewater treatment capacity and chemicals, and sludge-disposal fees.

Wastewater Treatment. Once the water stream has left the process, it must be treated before discharge. Several treatment choices are available. They typically consist of the processes listed below. Proper selection and design of the system will ensure that current discharge requirements can be met.

- Wastewater collection—waste streams from chrome-plating and cyanide baths are isolated and directed to the appropriate waste treatment unit. Effluents from these units are then combined in an averaging tank with other metal-bearing wastewaters, such as those from acid/alkali baths and rinses, other plating baths and rinses, and chemical dumps. They are then sent to the neutralization tank.
- Chromium reduction (as needed)—hexavalent chromium is converted (reduced) to the trivalent state, then precipitated as chromium hydroxide by alkali neutralization. A substitute process for chromium reduction is electrochemical reduction.
- Cyanide oxidation (as needed)—toxic cyanide-bearing waste streams are oxidized by chlorination or ozonation, forming less harmful carbon and nitrogen compounds.
- Neutralization/precipitation—the combined waste streams are treated with acid or alkali to adjust the pH to acceptable discharge limits and precipitate the dissolved heavy metals as metal hydroxides.
- Clarification—the neutralized waste stream is treated with coagulants and flocculants to promote the precipitation and settling of the metal hydroxide sludge, which are separated from the clarified liquid.
- Sludge handling—the collected hydroxide sludge is gravity thickened, mechanically dewatered, and sent to an approved hazardous waste disposal site.

While these processes effectively treat most heavy-metal-bearing waste streams, they may not be suitable for all applications. Furthermore, there is no guarantee that the "normal" design parameters (such as retention time and reagent dosage) will effectively remove the pollutants from every wastewater discharge. Treatability studies are often needed to verify the applicability of a treatment process to a specific wastewater.

Several alternative treatment processes have been developed to overcome the problems encountered in conventional treatment. Attention has focused largely on a problem encountered in the neutralization/precipitation step, wherein the solubility of dissolved metals cannot be brought to the low levels required for discharge. The problem arises when wastewater contains substances, known as chelating agents, that react with dissolved metals and interfere with their precipitation as metal hydroxides. Chelating agents such as ammonia, phosphates, tartrates, and ethylenediaminetetraacetic acid dihydrate (EDTA) are commonly used in plating operations and consequently find their way into the wastewater. Chelating agents react with the dissolved metal ion to form a "chelate complex" that is typically soluble in neutral or slightly alkaline solutions.

Two methods of overcoming the solubilizing effects of chelating agents are

- Precipitating the metal from solution by a method that, unlike hydroxide precipitation, is relatively immune to chelating effects; and
- Pretreating the wastewater to free the metal ion from the chelating agents.

The first category includes such processes as sulfide precipitation, ion exchange, and water-insoluble starch xanthate (ISX) precipitation. Sulfide precipitation, whereby metals are precipitated as sulfides instead of hydroxides, has achieved low levels of metal solubility in highly chelated waste streams. The process has been proven as an alternative to hydroxide precipitation or a method for further reducing the dissolved metal concentration in the effluent from a hydroxide precipitation system.

In ion exchange, a resin with a strong affinity for heavy metal ions (as opposed to the calcium and sodium ions normally present in the wastewater) is used to filter the ions out of solution. Ion exchange has proven to be a cost-effective means of lowering metal concentrations in electroplating discharges.

Water insoluble starch xanthate precipitation can remove heavy metal cations from wastewater. The ISX acts as an exchange material, replacing sodium or magnesium ions on the ISX surface with heavy metal ions in the solution. It is currently used both as an alternative to hydroxide precipitation and for "polishing" treated wastewater (to lower the residual metal concentration). Because ISX is insoluble in water and its precipitation reaction rate is rapid, it is used either as a slurry with the stream to be treated or a precoat on a filter through which the waste stream passes.

If the problems associated with chelating agents are not resolved by precipitation, the waste stream is typically segregated and pretreated either by raising the pH to a highly alkaline level (high pH lime treatment) or lowering it to an acidic level. At these extreme pH conditions, the metal complex often dissociates, freeing the metal ion. A suitable cation (such as calcium) is then used to tie up the chelating agent, preventing it from recombining with the metal ion when the solution is neutralized. While this type of treatment requires a high dosage of reagents, it has proved to be an effective means of treating some wastewater containing chelating agents.

Another frequent problem in wastewater treatment is metals concentrations in the effluent that exceed the discharge requirement even though the total amount of dissolved metals in the effluent is low. This condition indicates precipitated (undissolved) metals in the effluent, which results from an overloaded clarifier, ineffective conditioning (coagulation or flocculation), or poor pH control. The problem can be resolved by correcting the process deficiency or using a solids-removal (polishing) device, such as a sand or mixed-media filter, to clean the clarifier overflow.

BASIC DESIGN CONCEPTS

HYDROXIDE PRECIPITATION/COAGULATION. The conventional method of removing heavy metals is chemical precipitation of the metal as hydroxides followed by coagulation of the metal particles into larger, heavier floc particles, which then separate from the water (Metcalf and Eddy, Inc., 1991; Sittig, 1978; and Patterson, 1975). This frequently used method has proven reliable and can be inexpensive and highly selective. Generally, a properly designed and operated plant can reduce metal concentrations to 0.3 to 1.5 mg/L. In some cases, an even lower metal concentration may be achieved.

Typically, heavy metals are dissolved under acid conditions and precipitated under alkaline conditions. Thus, increasing the pH of a metal-containing solution to the alkaline pH range should induce the precipitation of the previously dissolved metal. Metals may redissolve at high pHs. The pH is increased by adding a base such as sodium hydroxide (NaOH, caustic) or calcium hydroxide ($Ca[OH]_2$, lime) to provide hydroxide ions. The heavy metal ions in solution react with the hydroxide ions to form solid particles. For example, with copper,

$$Cu^{+2} + 2NaOH \leftrightarrows Cu(OH)_2 \text{ (solid)} + 2Na^+$$

$$Cu^{+2} + Ca(OH)_2 \leftrightarrows Cu(OH)_2 \text{ (solid)} + Ca^{+2}$$

Though the pH at which a particular metal is least soluble (that is, most likely to precipitate) is a characteristic of that metal, this specific pH will vary depending on other components in the solution (such as chelating agents and surfactants) and conditions such as temperature. For example, if copper is the only cation in a deionized water solution, its minimum solubility, theoretically, is at pH 8.9. At pH above 8.9, all copper will be in solution. Below pH 8.9, copper will precipitate increasingly as pH decreases. Minimum solubility connotes the point at which no more copper will precipitate, not the point at which all copper will precipitate.

On the other hand, copper in a solution containing a mixture of copper, zinc, chromium, nickel, and iron might have a minimum solubility at pH 9.5. In this same solution, however, zinc might be least soluble at pH 8.5, chromium and nickel at pH 10, and iron at pH 9. Note that chromium precipitates as a hydroxide only if present in the trivalent (+3) state. Hexavalent (+6) chromium must first be reduced to trivalent chromium before hydroxide precipitation. Further explanation is provided later in this chapter.

Some laboratory experimentation is necessary to determine both the pH at which most metals are least soluble and the probability that the resulting concentrations are low enough to satisfy the regulatory limits. All wastewater, even to some degree those streams generated by similar industrial processes, may have a unique character that makes experimentation mandatory.

Figure 8.1 is a graph of pH versus copper concentration for the precipitation of a metal-finishing wastewater containing copper (initially 78 mg/L), zinc, chromium, nickel, and iron. The pH is adjusted to various values using lime, the solids are coagulated and settled, and the total and dissolved copper concentrations in the supernatant are analyzed. As the pH increases from 8 to 10.5, the concentration of copper decreases. In this sample, the lowest values result at pH 10.5, rather than at the theoretical point 8.9, and yet some of the copper is still dissolved because of chemical equilibrium. For complying with some discharge limits, the metal concentrations achieved at pH 9 to 9.5 may be adequate. The variability between wastewater types is evident in Figures 8.2 and 8.3, comparisons of pH versus zinc and nickel concentrations, respectively, in supernatants.

Once the appropriate pH is chosen and the metal hydroxides have precipitated, the next step is to coagulate the tiny particles into larger, heavier conglomerates or flocs. The heavier the flocs, the quicker they settle in a clarifier. Often, the hydroxide precipitates tend to floc together naturally, but not enough to remove all precipitates from suspension. Thus, a coagulant/flocculent aid may be added to enhance flocculation of particles, improving sedimentation and, ultimately, reducing heavy metal concentrations. This process typically is accomplished using commercially available polyelectrolytes; it is discussed in Chapter 5.

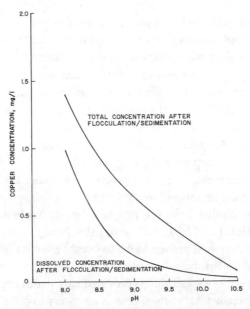

Figure 8.1   Total and dissolved copper concentrations in supernatant versus pH for an electroplating and electroless plating wastewater. Initial conditions: pH = 2.5, Cu = 78 mg/L, Zn = 0.53 mg/L, Cr = 79.5 mg/L, Ni = 96 mg/L, and Fe = 13.4 mg/L.

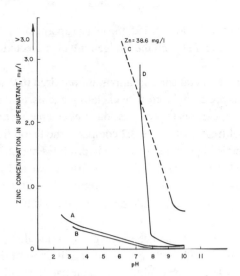

Figure 8.2 Comparison of pH versus total zinc concentration in supernatant for various wastewaters: (A) an electroplating and electroless plating wastewater containing Cr, Cu, Fe, and Ni; (B) a porcelain enamel wastewater containing Cr, Cu, Fe, and Ni; (C) an electroplating and metal-finishing wastewater containing Cd, Cr, Fe, Ni, and PO_4; and (D) a porcelain and enamel metal-finishing wastewater containing F, Fe, Ni, and PO_4.

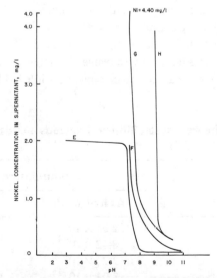

Figure 8.3 Comparison of pH versus total nickel concentration in supernatant for various wastewaters: (E) a porcelain enamel wastewater containing Cr, Cu, Fe, and Ni; (F) a porcelain enamel and metal-finishing wastewater containing F, Fe, Ni, and PO_4; (G) a metal-pickling wastewater containing Cr, Cu, Fe, and Zn; and (H) a metal-finishing wastewater containing Cr, Fe, Ni, Zn, and PO_4.

After the solids have settled, the clear supernatant may be discharged, if the pH is not too high, and the sludge may be collected for dewatering and final disposal.

When lower metal concentrations are required in wastewater discharges or if the metals are complexed with chelating agents such as cyanide or ammonia, sulfide or carbonate precipitation may be an effective associate or alternate form of treatment. Table 8.2 compares the theoretical solubilities of various metals in pure water as both hydroxides and sulfides. It can be seen that metals have lower solubilities as sulfides than as hydroxides, particularly in the neutral and alkaline ranges. Additionally, metal hydroxides tend to redissolve as pH increases, while metal sulfides continue to precipitate with increasing pH.

Optimum pH levels and chemical dosages should be determined in treatability tests along with settling tests to establish appropriate design rates for liquid–solids separation equipment. Precipitation should be carried out under conditions that allow the largely granular precipitate to grow to sufficient size for settling or rapid filtration. The resulting supernatants should be analyzed for heavy metals and compared for the best overall removal. Settling tests for clarifier sizing may be necessary.

Typically, the precipitates are filtered through a sand, paper tape, or other type filter. When ferrous sulfide is used, hexavalent chromium (Cr^{+6}) is reduced to trivalent chromium (Cr^{+3}) and precipitated as a hydroxide at pH 8.0 to 9.0; under such conditions, waters containing Cr^{+6} do not require segregation for individual treatment.

Sodium bicarbonate is not as effective at removing metals from solution as some other bases but has two advantages that must be considered. One advantage is that the metals can be precipitated while holding the pH within a

Table 8.2 Theoretical solubilities of various metals in pure water (pH = 7.0).

Metal	Solubilities, mg/L	
	As hydroxide	As sulfide
Cadmium	2.35×10^{-5}	6.73×10^{-10}
Chromium (Cr^{+3})	8.42×10^{-4}	No precipitate
Copper	2.24×10^{-2}	5.83×10^{-18}
Iron	8.91×10^{-1}	3.43×10^{-5}
Lead	4.02×10^{-3}	5.48×10^{-10}
Nickel	6.92×10^{-3}	6.90×10^{-8}
Silver	13.3	7.42×10^{-12}
Zinc	1.1	2.31×10^{-7}

narrow range at nearly optimum levels and maintaining the alkalinity of the wastewater. This, in turn, permits a simpler control system design. Another advantage is the neutralization of excess activity (that is, it adds buffering capacity), thus helping to meet discharge standards.

A bicarbonate–carbonate mixture can precipitate more metals than bicarbonate alone. Sodium bicarbonate can only bring the pH of a solution to approximately 8.3, which is not high enough to produce, through precipitation, effluent below regulatory limits for metals such as nickel and cadmium. An increase in the pH of a solution to 9.0+ will cause precipitation of additional metals. A carbonate must be combined with the bicarbonate because of its buffering effect to reach this pH.

The required pH and proper carbonate alkalinity can be tested and determined for the metals involved. The distributions of the three forms of carbonate (CO_3^{-2}, HCO_3^-, H_2CO_3) remain the same for any particular pH level. Treatment efficiency can be increased by adding more carbonate. This presents an advantage over using sodium hydroxide as the base, because adding more sodium hydroxide above a certain level will not remove additional metals (that is, an overdose of sodium hydroxide could move the process pH to a region at which metal solubility increases because of metal hydroxo complex formation).

The use of sulfides in metal precipitation requires more care than the use of a hydroxide form. Excess sulfides in an alkaline solution will form hydrogen sulfide (H_2S), an objectionable odorous gas that is lethal at low concentrations in confined spaces. In an acidic solution, sulfides can produce larger quantities of H_2S, which can be poisonous and typically more pungent. In addition, the cost of sulfide precipitation typically is higher than that of hydroxide precipitation. Sulfide treatment also requires continuous operator attention to control the toxicity hazards associated with sulfide. Finally, the process may result in larger sludge quantities, and the disposal of metallic sulfide sludge may pose environmental risks.

Two distinct sulfide precipitation processes, insoluble and soluble, are used to treat wastewaters containing heavy metals. Insoluble sulfide precipitation (ISP) uses ferrous sulfide as the sulfide source. Ferrous sulfide is relatively insoluble in water; consequently, the level of dissolved sulfide in the wastewater is kept to a minimum. The main advantage of ISP over soluble sulfide is that no detectable hydrogen sulfide odor is associated with the process. Soluble sulfide precipitation uses a water-soluble reagent such as sodium hydrosulfide or sodium sulfide (Na_2S).

Figures 8.4 and 8.5 show the relative solubilities of various metals as hydroxides and sulfides, respectively, versus the pH of the solutions.

Low-level Metals. Hard-to-precipitate low levels of metals such as molybdenum can sometimes be coprecipitated with other insolubles. One mining company trying to lower trace parts per million levels of molybdenum was able

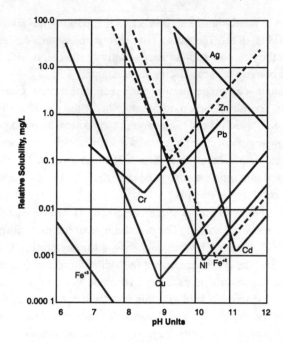

Figure 8.4 **Relative solubilities of metal hydroxides versus pH (U.S. EPA, 1980).**

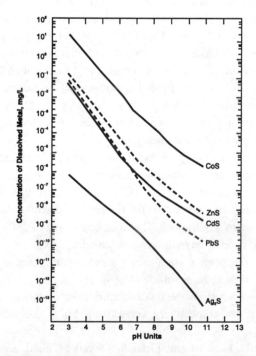

Figure 8.5 **Relative solubilities of metal sulfides versus pH (U.S. EPA, 1980).**

to get the molybdenum sorbed onto insoluble colloidal manganese dioxide and then filter the manganese dioxide.

APPROACH TO DESIGN. Data must first be collected to define the nature and extent of the problem and identify the most appropriate solution. The designer should be familiar with the industrial processes generating wastewater and the plant layout. To define the nature and extent of the problem, the designer should collect wastewater samples and flow measurements at three levels: continuous and noncontinuous flows, individual (for example, a rinse tank or a batch process) and end-of-pipe (all wastes from the plant) discharges, and individual tanks that occasionally are dumped. The samples should be analyzed for suspected contaminants that could interfere with precipitation and sedimentation (for example, oil and grease or surfactants), and contaminants for which limits have been prescribed. Once the contaminants have been identified, the designer can develop treatability studies for optimum treatment of the wastewater and sludge, design criteria for sizing alternative facilities, and guidelines for pinpointing potential problem areas (such as batch dumps) that require special handling. These alternatives may be compared further on the basis of initial capital investment, operation and maintenance costs, project life, space requirements, and reliability. Pollution prevention and waste minimization practices should be evaluated early in the waste management strategy development.

Familiarization with, and an understanding of, the various industrial processes in a plant are essential to designing the sampling program and analytical scheme. For example, for a pickling process in a metal-finishing plant, the necessary data are the sequence of the process, tank sizes, rinse flows, chemical makeup of the concentrated baths, temperature, dump schedule for spent baths, drag-out, and hours of operation. Surface area or weight of the material processed is also helpful, especially in light of existing and future regulations that compare waste load to processed finished product. These data eventually will assist in sizing treatment facilities, evaluating water-use-reduction potential, determining sources of specific metals or interfering substances, and selecting equipment construction materials.

After the wastewater sources are identified, the sampling program may be established. Obtaining representative samples requires that sampling continue for two or three plant operating periods, at a minimum, at those plants where processes are consistent from day to day. Industries with frequently varying processes should tailor the sampling program to their unique schedules.

The treatability study initially is based on analytical results and flow determinations. Sometimes, because of the plant layout, old piping connections, and the practical manner in which sampling has to be carried out, samples must be mixed together in flow-proportional composites to simulate end-of-pipe wastewater. Also, if water-use reduction is conceivable, a composite should be mixed to simulate how this change will affect the treatment scheme. It may also be necessary to obtain additional separate samples for pretreatment studies of cyanide or hexavalent chromium reductions and concentrated metals precipitation. Decreasing the amount of water increases the metals concentrations; therefore, more precipitates form per volume of water. The more solids present, the more likely they are to flocculate and thereby enhance settling and metals removal. Two other benefits of decreasing water use are lower monthly water bills and smaller pretreatment equipment size and cost.

After mixing the composite that simulates the future pretreatment influent, initial metals concentrations and pH should be measured. A series of tests may then be conducted to vary the precipitation reaction pH. After the precipitates settle out, the metal concentrations of the supernatants are measured, and the pH at which all of the metals have the lowest concentrations, or at which they are within the required limits, is chosen as the optimum reaction pH. This evaluation typically is conducted with lime or sodium hydroxide (NaOH), but other alkaline chemicals also may be used, depending on the circumstances. From a maintenance viewpoint, NaOH is generally easier to add. Lime works better, however, when metal concentrations are dilute because it adds bulk to the solution and enhances sedimentation. Lime removes high concentrations of sulfate ions by precipitation of $CaSO_4$. Lime is considerably cheaper, but it produces more sludge, must be made into a slurry and pumped, and obstructs piping. When these chemicals are added to wastewater, it is important to note the volume required to increase the pH to the desired value. This information is necessary for sizing such items as pumps and chemical storage facilities and estimating operational expenses.

If the solids precipitated by pH adjustment settle rapidly and leave a clear supernatant, polymer addition may not be necessary. Usually, however, the next step in the treatability study is jar testing—testing to determine the best polymer and its optimum dosage. A series of jars or beakers containing pH-adjusted wastewater samples is lined up beneath a series of mixers. To each jar is added simultaneously either a different polymer or a different dosage of the same polymer. The samples are then mixed for a predetermined time. In this manner, the efficiency of polymers or dosages may be visually compared. Those resulting supernatants with the best appearance may be compared analytically. The desirable performance characteristics of a polymer are the production of a firm floc particle with rapid settling qualities and generation of a clear supernatant with few particles (fines) remaining in suspension. In addition, the speed of the mixer may be varied to determine the optimum energy input requirement.

After the best polymer at its optimum dosage is selected, a settling test is conducted. First, the pH-adjusted water is added to a graduated cylinder or a container designed for settling tests. The polymer is added at the established dosage and mixed. Immediately after turning off the mixer, the operator records the height of the sludge interface versus time until the settling rate approaches 0. These values are then graphed to produce the characteristic settling curve. This curve is used to determine the overflow rate for a sedimentation process (Metcalf and Eddy, Inc., 1991).

Metal hydroxides coagulated with a good polymer at the proper dosage typically settle quickly. However, because of potential plant upset conditions, a safety factor should be applied to the calculated overflow rate. This overflow rate is used to size the effective surface area of the sedimentation process. For a thick floc that settles immediately, the calculated overflow rate could be as high as 60 m^3/m^2·d (1 500 gpd/sq ft). However, the design overflow rate typically should not exceed 40 m^3/m^2·d (1 000 gpd/sq ft). More often, the values range between 10 and 33 m^3/m^2·d (250 and 800 gpd/sq ft).

With the common use of chelating agents in production processes, it is often found during treatability studies that some metals remain in solution throughout the treatment process. Consequently, the destruction of these chelating agents may be necessary during the pH adjustment process or aside from it. The advent of these types of agents almost mandates treatability studies unless historical treatment data is available on the particular waste.

The final step in the treatability study is to evaluate sludge dewatering capability. Most hydroxide sludge after settling is 96 to 99% water and typically may be dewatered to 65 to 85% water. Currently, the most common methods for hydroxide sludge dewatering are sand drying beds, vacuum filters, or filter presses. Sand drying bed performance is climate dependent but may be partially simulated in the laboratory by setting up a sand column, applying sludge to the surface, and monitoring the filtrate quantity, quality, and final sludge moisture content. Plate and frame pressure filters or belt filter presses are used with most new systems, with sand and vacuum filters declining in use. Filter press applications may be designed from bench-scale tests that have been developed by various manufacturers.

The specific resistance test describes the filterability of the sludge. Typically, the sludge must be chemically conditioned with a compound such as lime, polymer, or ferric chloride before filtering. This test aids in determining what chemical to use and the optimum dosage for filtering particular sludges. The optimum dosage results in the lowest specific resistance value. A sample of conditioned or unconditioned sludge is applied to a Buchner funnel apparatus under vacuum, and the quantity of filtrate is monitored versus time. The initial and final sludge moisture contents also are determined. A graph is constructed by plotting filtrate volume versus time divided by volume, and the slope b of the best-fit line through the points is calculated. The specific resistance value r, in centimetres per gram, is calculated by the formula

$$r = 2bPA^2/VC \qquad (8.1)$$

Where

b	=	slope of the filtrate curve, s/cm^6;
P	=	applied vacuum, kN/m^2;
A	=	area, cm^2;
V	=	filtrate viscosity, N s/m^2; and
C	=	solids deposited per unit volume of filtrate, g/cm^3.

The results of the data collection effort will determine the necessary unit processes and their sizes in the final pretreatment design. A generalized schematic of a metal hydroxide precipitation/coagulation facility is shown in Figure 8.6. Numerous specific designs fit within this general framework. Equalization is necessary to dampen flow, pH, and concentration fluctuations so that a consistent wastewater stream enters the critically controlled unit processes. The pH-adjustment step should occur in a mixed tank of about 10 to 15 minutes detention. Design should follow proper engineering practices and address specific metals-related needs.

Polymer addition also may take place in a mixed tank of about 10 to 15 minutes detention, or it may be fed in the line to the solids-separation process. This choice typically depends on the solids-separation process chosen.

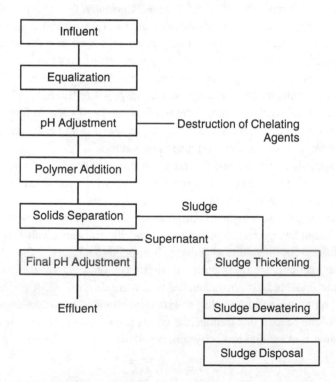

Figure 8.6 Generalized schematic of metal hydroxide precipitation/coagulation facility unit processes.

These processes are described for metals removal in the next section. One should also refer to Chapter 5 for general design assistance.

The supernatant may require final pH adjustment before discharge to the sewer if the pH end-point decided on is higher than discharge limits allow. Occasionally, discharge limits for heavy metals are lower than those that hydroxide precipitation can accomplish. In such a case, another method must be investigated for primary removal or as an add-on, polishing unit process—a method, such as sulfide or carbonate precipitation or filtration, that uses sand, multimedia, or paper.

The sludge produced is collected for further dewatering. It may then be applied directly to a sand drying bed, from which it must eventually be removed and transferred to appropriate disposal. The hydroxide sludge typically may be further thickened before being dewatered mechanically. The cake produced may be good landfill material in terms of moisture content; however, it must be evaluated under the Resource Conservation and Recovery Act (RCRA) regulations for hazardous waste classification. By definition, metal-plating sludge is a RCRA-listed hazardous waste, F006, and may not be transported to a nonhazardous waste permitted landfill.

SOLIDS-SEPARATION PROCESS OPTIONS. There are many possible ways to separate from the water phase the solids produced during precipitation/coagulation. The ones most commonly used in metal hydroxide treatment are described briefly, and schematic drawings of examples of each are shown in Figure 8.7.

Sedimentation Pond. A sedimentation pond or lagoon is a lined or concrete, and typically rectangular, basin in the ground. Its required size may be determined in one of two ways. First, it may be designed based on the overflow rate calculated in the treatability study, to provide a required surface area. It should be of adequate dimensions to ensure that the smallest particle characteristic of that wastewater has sufficient time in which to settle. As sludge settles in the bottom of such a basin, the efficiency of the process diminishes and the accumulated sludge must be removed.

Alternatively, the basin may be designed on the basis of sludge accumulation. For example, it may be designed with more than one cell: while one cell is being cleaned, the other can be placed in operation. One disadvantage of sedimentation ponds, however, derives from the inherent nature of hydroxide sludge, which is hydrophilic and, thus, does not dewater easily over time. Rather, the particles tend to bridge together. Sedimentation ponds also require large areas of land.

Conventional Clarifier. The conventional clarifier was developed to overcome the problems encountered with sedimentation ponds. Clarifiers may be rectangular or round, with concave, cone-shaped bottoms. They are equipped

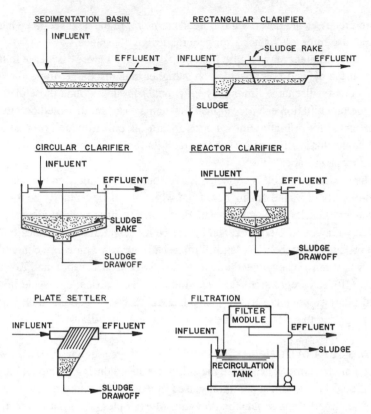

Figure 8.7   Schematic drawings of examples of solids separation processes used in metals removal.

with power-driven, rotating rake mechanisms that gather settled sludge on the bottom to the low point of the cone for draw-off. A rectangular clarifier or sedimentation basin may be equipped with a traveling sludge collector that continuously scrapes the sludge to a hopper.

Clarifier size is based on the design overflow rate for the effective surface area. The sludge is periodically transferred to a holding tank or lagoon. Sometimes, the sludge will dewater and thicken further while being retained. A holding tank design that calls for sludge filtration or mechanical dewatering is called a sludge thickener. It is similar to a clarifier but sized according to volume based on detention time before dewatering.

Solids Contact Clarifier. A solids contact clarifier is similar to a conventional clarifier but with a central, inverted cone near the top. The pH-adjusted wastewater, along with the polymer, enters the clarifier through the top of the cone. Inside the cone, the floc forms and begins to settle out. Instead of continuously being drawn off, the sludge is allowed to accumulate to a level above the lower edge of the cone. This thick layer of sludge, called a "sludge blanket," serves as a nucleus for floc formation and a filter for the smaller floc particles. The water flows through the blanket and up over the weir around the top of the clarifier.

Plate Settler. A plate settler is a deep, rectangular unit equipped with several parallel plates slanted at an angle (often 45 or 55 deg) and spaced 25 to 50 mm (1 to 2 in.) apart. A commonly used system is the lamella settler. The pH-adjusted, polymer-treated wastewater flows upward through the plates. Floc particles settle onto the plates and slide into the sludge-holding area of the unit. The advantage of this unit is that the flocs have a shorter distance over which to settle, and an effective laminar flow condition may more easily be attained to increase settling. In addition, because a large surface area is provided by the closely compacted slanted plates, the overall unit size is smaller than a conventional clarifier of the same effective throughput. Furthermore, the unit may be manufactured elsewhere and delivered to the construction site for installation.

Under heavy loadings, operation problems may develop, causing solids to bridge in the sludge-holding area and become difficult to remove. This problem is compounded by excessive polymer addition. Angled plates may also be installed in conventional clarifiers to increase the overall surface area.

Filtration Systems. Filtration systems are designed to eliminate the sedimentation step. For example, once the floc has formed, the floc and wastewater can be filtered through a filter press (often a diatomaceous earth precoat is used), resulting in a filtrate and sludge cake. Another unit currently available consists of a filter module installed over a storage tank; the sludge is concentrated in the filter module and directed to the storage tank. If the filtrate is clear, the sludge may be discharged; if not, the filtrate is treated further, or the sludge is recirculated to the process.

OPERATIONAL CONSIDERATIONS. Frequent checks of the pH-adjustment system are necessary; pH probe cleaning and periodic calibration (checking to see whether the pH reading is correct) are advisable because hydroxide precipitation is primarily pH dependent.

Frequent jar tests are required to increase polymer dosage and effectiveness, especially for systems without equalizations.

The hydroxide precipitation/coagulation treatment process is effort intensive. Much time is spent in the sludge dewatering process cleaning the filters or removing dried cake from sand beds; preventing stoppages in sludge-transfer plumbing; preparing chemical solutions; maintaining pumps, valves, mixers, sludge rakes, and controls; and performing chemical analyses to evaluate efficiency.

A major disadvantage of the neutralization/precipitation process (as the pretreatment limits of all metals are reduced) is that the typical metal-finishing industry is characterized by a number of lines using different metals

---

and batch dumps, resulting in a widely varying waste stream. Thus, selection of an optimum pH value to achieve minimal metal concentrations, even with equalization, is difficult and may change occasionally. Also, large hazardous sludge volumes are produced, with costs of disposal increasing.

Thus, many metal finishers (both platers and printed-circuit board manufacturers) are increasingly turning away from this treatment to ion exchange, ultrafiltration, or evaporation, which all produce lower effluent metal concentrations and sludge volumes and produce high-quality water for reuse.

The metal-finishing industry has been targeted by regulators as a major hazardous waste generator, making conventional neutralization processes less desirable. These new processes, and their ability to generate reusable water, are discussed later.

CHEMICAL CONVERSION. There are two types of chemical conversion typically required before hydroxide precipitation: hexavalent chromium reduction and cyanide destruction. These conversions are best performed in segregated treatment units so that a smaller volume of wastewater is treated, rather than the entire pretreatment effluent. Because hexavalent chromium and cyanide must be converted, many plants avoid using these substances whenever possible.

Hexavalent Chromium Reduction. In some industries, chromium is used in the hexavalent (+6 valence state) rather than trivalent (+3 valence state) form because of certain preferred qualities in the manufacturing process. Although Cr^{+3} will form $Cr(OH)_3$, a hydroxide precipitate, hexavalent chromium (Cr^{+6}) persists as an ion even when the pH is increased by the addition of a base. Hexavalent chromium cannot be removed as a hydroxide. However, Cr^{+6} may be converted to Cr^{+3} by chemical reduction. This reaction occurs quickly at pH 2 and involves the addition of SO_2 (sulfur dioxide), sodium bisulfite, or sodium sulfite, until the reaction is complete. The completion of the reaction may be monitored by an oxidation-reduction potential (ORP) meter. The ORP reading may vary with wastewater characteristics. A color change from yellow to green is also evident. At this stage, base may be added to precipitate the Cr^{+3}, or the wastewater may be added to other plant wastewaters for precipitation.

Cyanide Destruction. Cyanide (CN^-) destruction is important in metal removal because cyanide forms complexes with metals and prevents them from precipitating as hydroxides. Once the cyanide–metal bond is broken, however, the metal is free to precipitate under the appropriate pH conditions. Cyanide destruction is also important in pretreatment because slug doses of cyanide to a municipal biological system may destroy the active organisms or pass through the plant to the receiving waters. Low concentrations of cyanide (free or complexed forms of the CN⁻ ion) may be destroyed by treatment plant organisms acclimated to it.

Cyanide destruction typically is a two-step process. The first step involves converting cyanide to cyanate with sodium hypochlorite at pH 10 or greater. Decreasing the pH with acid to 8.5 converts cyanate to carbon dioxide and nitrogen. The first step requires approximately 30 minutes. The reaction endpoint may be monitored by an ORP meter and a color change from green to blue. The second step requires about 10 minutes. For design purposes, the exact amount of chemicals needed for these reactions should be determined by experimentation. In operation, the reactions may be controlled by feedback ORP and pH meters.

The resulting wastewater can be added to other plant wastewaters for metals precipitation. If cyanide is widely used in the manufacturing process, an evaporative recovery system may be attractive for the purpose of recycling some cyanide and concentrating waste cyanide in such a manner that destruction facilities are smaller in size and cost. Other methods for cyanide destruction include ozone treatment, heat/pressure, electrolysis, and hydrogen peroxide.

OTHER DESIGN CONCEPTS

Some other heavy metal removal methods are discussed in the following paragraphs. Some of these may be used as end-of-pipe treatment schemes, but economically they are more suitable either as polishing treatments (sulfide or carbonate precipitation, for example) or in-line conservation techniques (such as ion exchange or reverse osmosis). Others are still in the development phase and unproven in full-scale applications. All are given increasing consideration in meeting pretreatment requirements.

ION EXCHANGE. The ion-exchange process entails the exchange of desirable ions from the resin for undesirable ions of similar electrostatic charge dissolved in wastewater. The resin has a cross-linked structure of particles that adsorb specifically charged ions. The cation resin in its unused state has cations, such as H^+, attached to the resin particles. As the metal-containing wastewater is in contact with the resin, the metal ions exchange adsorption sites with the H^+ ions. The resin then has metal ions attached to it, and the wastewater contains H^+ ions. Eventually, most of the sites will have heavy metals attached. The cation resin must then be regenerated with acid solution to remove the metals and reactivate the resin with H^+ ions. This solution containing the metals will require treatment, such as hydroxide precipitation, for metal removal or reuse (electrowinning).

Some platers have installed ion-exchange processes to treat wastewaters discharged from plating baths. The acidic regenerant is used to make up the plating bath, with virtually no waste requiring treatment. Similarly, anion resins can exchange sulfates, chlorides, chromates, and cyanides for more desirable ions such as OH^-. A variety of resins are available for specific applications, each having a preference for certain ions.

In wastewater treatment, the resin typically is contained within a column, and the wastewater is passed through the column under pressure. Thus, for an installation of four columns, two cation resin-filled columns and two anion resin-filled columns, the operation may occur in series or parallel. Parallel operation is simpler. As the resin in the first column of the cation columns becomes exhausted (determined by the conductivity of the effluent), the wastewater flow is diverted to the second column while the exhausted column is regenerated. In series operation, wastewater flows through both columns, typically with the second column acting as a safeguard so that no leakage occurs when the first column is exhausted. At that time, the influent is diverted to the second column, and the first column is regenerated and put back into service as the second column. This system requires more operator attention but fewer regenerating chemicals.

Regeneration of the columns is accomplished when using the following process. First, the column is drained of wastewater and the resin is agitated by a combination of compressed air, followed by backwashing with potable water and draining. After ventilation releases the compressed air, regeneration occurs. Acid (8 to 10% by volume concentration) regenerates the cation resin, and caustic (4% concentration) regenerates the anion resin. The chemicals are drained, potable water is added, and rinsing continues until the proper pH is attained (pH 3 to 4 in cation column effluent, pH 10 to 11 in anion column effluent). The column is then ready for service. All water and chemicals used in regeneration must be treated for metal removal.

Ion exchange effectively removes metals. However, because of its initial cost, operation and maintenance requirements, and chemical regeneration costs, ion exchange is more applicable to certain situations. To preserve the life of the resin and reduce chemical costs, the ion-exchange system should be designed so that regeneration is required not more than once per day, and preferably not more than once per week. If the wastewater influent to the columns is concentrated, the resin will be exhausted quickly and require frequent regeneration. Conversely, because column size depends on hydraulic load, reducing the flow to the columns will reduce the initial cost of the system. Balancing these two aspects—reducing chemical as well as hydraulic load—is the key to a successful, economical ion-exchange system.

Where effluent standards are extremely strict, ion exchange has been used as a polishing process after hydroxide precipitation/coagulation. This end-of-pipe solution may be impractical, however, because the columns must be sized for the total plant flow, though the chemical load should be low.

Because ion-exchange effluent typically is of better quality than the plant's potable water, the process also may be used to provide rinse water to manufacturing processes. Chemical and hydraulic loads to the ion-exchange system may be balanced by treating only the rinse water from the industry's processes. The most commonly used resins for this application are strong acid cation resin, weak basic anion resin, and strong basic anion resin (for CN^- and F^- removal). All other wastewater, along with regenerant waste, would be treated in a smaller and less expensive conventional metal-removal system. In addition, the industry saves money by recycling—a welcome bonus in light of rising water prices. This solution is particularly attractive to the metal-finishing and printed-circuit-board industries.

Ion exchange may be used to recover valuable substances at their sources. Many photographic labs recover silver by passing the film-developing wastewater through ion-exchange columns and collecting the silver in the regenerating solution. Recovery of other noble metals, such as gold and platinum, is also common. As other metals like copper increase in value, ion exchange will become more economically attractive.

To design an ion-exchange system, the engineer must determine the chemical and hydraulic loads of the column influent. For end-of-pipe polishing, the contaminants in the effluent from the primary metal-removal treatment must be analyzed to calculate the chemical load. Some bench testing of resin-filled columns also may be necessary. The hydraulic flow determines the column size; the chemical load determines the quantity of resin, service time, and quantity of chemicals for regeneration.

For ion-exchange recycling in manufacturing processes, detailed data should be collected from rinse waters on drag-out and chemical load. For example, for metal finishing, data should include the sequence of steps within the process; tank sizes; chemical makeup of and periodic additions to concentrated baths; rinse flows; rinse water temperature and quality requirements; presence or absence of stagnant rinses; and dumping schedule.

Drag-out analysis must be performed to determine the quantity of water transferred from tank to tank while a known quantity of material is processed over time. Using this information, the engineer calculates the average chemical load in each rinse, necessary minimum rinse flow to maintain a desired concentration, and the necessity for, and volume of, stagnant rinses to decrease the chemical load of the flowing rinses. The final results include the size of the ion-exchange system, its service time between regenerations, and the volume of chemicals for regeneration. Moreover, the treatability of the wastewater for the accompanying precipitation system must be evaluated. With the rinse water no longer available for dilution, the precipitation system will treat more concentrated and potentially problematic wastewater. Treatability must be carefully evaluated through bench testing, as described previously.

Finally, the combination of an ion-exchange system for recycling and a conventional system to treat the other wastewaters should be assessed in terms of overall system cost and operation and maintenance costs—water savings, space requirements, future goals of the industry, and level of operator training.

Ion exchange demands the attention of an operator who understands the sophisticated process and its controls. For a combination recycling ion-exchange and metal-precipitation system, 20% of the operation and maintenance time typically is devoted to ion exchange and 80% to precipitation.

The resin-filled columns must be preceded by a multimedia filter with activated carbon to prevent the resin from becoming clogged with suspended solids and fouled with organic material. The backwash from the filter also must be treated.

In a recycle system, special care should be taken to coordinate the manufacturing process with the ion-exchange system because the two ultimately will be closely interrelated. Periodic blowdown and makeup water addition will be necessary.

ADSORPTION. Adsorption is the adhesion of dissolved substances to the surface of solid particles. Several adsorption processes for metal removal are in the development phases. Few have proven successful in full-scale applications for waste treatment, though much bench testing and pilot work have been conducted. A brief description of a few of these new technologies is provided in the following paragraphs.

Activated Carbon. Although activated carbon has been used in the removal of organic substances from wastewater for several years, its application in the area of metals removal is recent. The wastewater first must be treated by sedimentation or filtration to remove precipitated metals. The wastewater containing dissolved metals primarily is contacted with the activated carbon in a tank or column. The metals adhere to the carbon until all available sites are exhausted, at which time the spent carbon is replaced with new or regenerated carbon.

Soluble Sorbent Clarification. Soluble sorbent clarification is a proprietary process using a specific chemical sorbent that is soluble at low pH and insoluble at high pH. The pH of the wastewater containing the dissolved sorbent is increased to precipitate hydroxides. At the higher pH, the remaining dissolved metals adsorb to the now insoluble sorbent, causing it to settle with the hydroxides. The sorbent may be recovered by acidifying the sludge, thus solubilizing the sorbent at a low pH. Proponents of the process have reported experimental removals of copper, nickel, zinc, and cadmium to levels of 0.02 to 0.05 mg/L, and chromium to 0.1 mg/L.

MEMBRANE PROCESSES. Ultrafiltration. Ultrafiltration is one of several membrane separation processes. It is primarily a concentration process. A substantial amount of pumping energy is required to force liquid through the membrane while relatively large-sized particles (molecular weight 10 000 to 40 000 and greater) are retained and concentrated in the liquid portion, which does not pass through the membrane. This process has a potential use in the recycling of metal-containing alkaline cleaners and paints (the concentrate). The overall result is the reduction of metals in wastewater (the filtrate).

Reverse Osmosis. Reverse osmosis is also a membrane process. However, it concentrates dissolved materials from dilute solutions. Both the size of the ion and its charge tend to affect the portion that will pass through the membrane; thus, an absolute separation typically is not achieved. The small, single-charged hydrogen ion seems to pass readily through the membrane. Large metallic ions and those that carry a higher positive charge (such as nickel, zinc, copper, arsenic, and cadmium) do not pass through the membrane with the same freedom as do lower, positively charged ions (such as sodium and potassium).

A great deal of pumping energy is required to force the liquid through the membrane from the lesser to the more concentrated solution, because the pressure must overcome the osmotic pressure between the solutions—thus, the name "reverse osmosis." This system is not generally applicable to the total discharge. It typically is used nearer the source of a particular offending substance, either to recover or concentrate it to make chemical treatment easier.

Reverse osmosis has been used by a number of metal platers for controlling nickel in the rinse discharge following the nickel plate. It has been proposed by U.S. EPA as a means of "zero discharge" for this portion of the plating operation. Even these applications, however, have not been proven successful for a full-scale operation. Although reverse osmosis decreases the quantity of discharge, occasionally there must be a blowdown from the system to remove dissolved materials that might interfere with proper plating actions in the plating bath.

Electrodialysis. Electrodialysis makes use of the inability of some chemicals to pass through a charged membrane as easily as water. Electrodialysis reversal conducts small amounts of direct current electricity through the wastewater. The charged impurities move through alternating positive and negative resin membranes that remove ionic impurities from the water. Electrodialysis and electrodialysis reversal can remove ionic pollutants such as sulfates, fluorides, and chromates. Membrane fouling and scaling are potential problems.

EVAPORATION

Evaporation may be used to recover useful byproducts from a solution or concentrate wastes before additional treatment and disposal. Some evaporation applications may also result in the recovery of a pure solvent from solution.

During evaporation, a solution is concentrated when a portion of the solvent, typically water, is vaporized, leaving behind a concentrated, saline liquor that contains virtually all of the dissolved solids, or solute, from the original feed water. The evaporation process may be carried out naturally in solar evaporation ponds or via a commercial mechanical evaporator. Air emissions potential must be evaluated.

EVAPORATION PONDS. In locations where the annual evaporation rate exceeds the annual precipitation rate, solar evaporation ponds may be used to handle small or problem waste streams.

A solar evaporation pond is an open holding pond or lagoon that depends solely on climatic conditions, such as precipitation, temperature, humidity, and wind velocity, to effect the evaporation of a solvent (typically water) from a wastewater solution.

Ponds must be sized to handle maximum wastewater flows as well as the solids accumulation resulting from evaporation. An impervious pond liner, underground leachate collection system, and leak-monitoring system typically are required. Provisions must be made to periodically remove accumulations of solids for disposal and inspect and repair the pond liner. Wind erosion and wave action may occasionally require that the pond be rebuilt.

Application of evaporation ponds is usually limited by land cost and availability or climatological conditions.

MECHANICAL EVAPORATORS. The use of mechanical evaporators is an effective method of concentrating or removing salts, heavy metals, and other hazardous materials from solution. The evaporation process is driven by heat transferred from a condensing steam to a lower temperature solution across a metallic heat-transfer surface. The absorbed heat causes vaporization of the solvent and an increase in the solute concentration. The resulting vapor may be vented to the atmosphere or condensed for reuse.

Other methods of heating the water to promote vaporization include the use of hot oil, electric gas, fuel oil, waste heat from existing processes, and heat pumps.

Because evaporation is an energy- and capital-intensive treatment process, the selection and design of an evaporator system must be carefully considered for each application.

Evaporators have been used successfully in many industrial treatment applications, including

- Heavy metal waste streams;
- Electroless metal-plating waste streams;
- Emulsified oil streams;
- High soluble biochemical oxygen demand (sugar) streams; and
- Nonvolatile aqueous organic or inorganic streams (for example, dyes, acids, and bases).

Evaporators may be an attractive alternative in many of these applications because they can concentrate and recover valuable materials for reuse without chemical addition. Assuming proper construction materials are used, evaporators can be operated with virtually any combination of metals or nonvolatile organics at any metal concentration.

In many metal-finishing and metal-processing applications, evaporators are used to achieve zero liquid discharge of rinse water from the various manufacturing and coating processes.

Evaporation has a number of advantages over conventional physical–chemical treatment processes. One of the most significant advantages is the high-quality distillate produced (usually < 10 mg/L total dissolved solids). Most conventional treatment processes are not able to meet current and proposed NPDES discharge limits consistently without extensive effluent polishing, and the effluent from these processes typically is not acceptable for reuse within the plant.

Not only does effluent (distillate) produced by an evaporator meet most discharge limitations, it can almost always be recycled for reuse in a manufacturing process. In at least one metal-finishing application (Patterson, 1975), distillate was able to be recycled as process rinse water at 10% the volume of municipally supplied water required to do the same job.

In addition to producing a high-quality distillate, an evaporator reduces the production of regulated waste residues and increases the possibility of recovering valuable metals from those wastes. Evaporation is less sensitive to small amounts of oil than ion exchange and produces no regeneration wastes that require additional treatment.

One of the fastest growing areas in the metal-finishing industry is the use of the "electroless" plating process common in the manufacture of printed circuit boards. This process produces a hard-to-treat waste stream that contains chelating agents such as organic acids (that is, EDTA) or ammonia, which prevent the normal precipitation of heavy metal hydroxides. This necessitates the use of additional chemicals to "break" the chelating agents before conventional treatment. However, evaporation has been used successfully on electroless plating wastes without additional chemicals.

The concentration of liquids by mechanical evaporation is an operation that requires considerable quantities of energy. It is therefore necessary to consider various energy alternatives to select the most efficient type of evaporator.

In an ideal system, 1 kg of condensing steam will evaporate 1 kg of water from the solution. Such a system has a steam efficiency, or economy, of 1:1 (1 kg of water removed for every kilogram of steam supplied). A simple evaporator system has a single evaporation chamber, or effect, and is said to have an "economy of one."

Evaporator economy can be increased by increasing the number of effects. A multiple-effect evaporator uses the vapor from the first effect as the steam source for each subsequent effect, which boils at a slightly lower pressure and temperature (Figure 8.8).

Each additional effect increases the energy efficiency of the evaporator system. For example, a double-effect evaporator requires approximately 50% of the steam required by a single-effect unit and is said to have an economy of 2.

The number of effects can be increased to a point at which the capital cost of the next effect exceeds the savings in energy costs.

Vapor compression is another proven technique to reduce energy requirements. In a vapor compression system, vapor discharged from the evaporator chamber is compressed to the pressure/temperature required in the evaporator heat exchanger (Figure 8.9).

Mechanical compressors are the most frequently used method of vapor compression. Compressors may be of the positive displacement, centrifugal, or axial types. An evaporator system using mechanical vapor compression will require an outside steam source only to initiate operation of the unit.

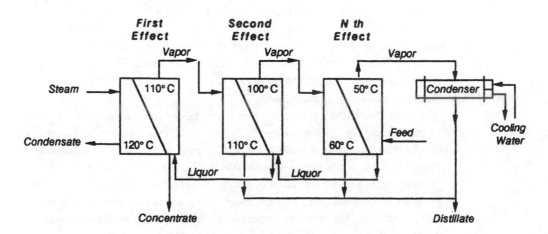

Figure 8.8 Multiple-effect evaporator.

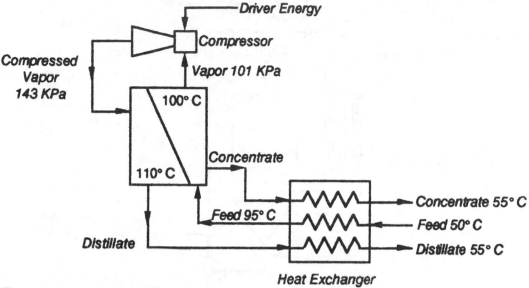

Figure 8.9 Vapor compression evaporator.

This typically can be supplied by a small boiler or resistance heater in the evaporator feed tank.

A steam jet thermal compressor using high-pressure steam to compress vapors to the required pressure can also be used. The use of a thermal compressor is approximately equivalent to adding an additional evaporator effect.

When available, waste steam or heat from other process streams may also be used to lower evaporation costs. Hot process fluids can be pumped through the heating tubes instead of steam, recovering heat and transferring it to the fluid to be evaporated.

Mechanical evaporators typically are categorized according to the arrangements of their heat transfer surfaces and the methods they use to impart energy (heat) to the solution.

Vertical Tube Falling Film. Recirculating liquor (process fluids) is introduced at the top of a vertical tube bundle and falls in a thin film down the inside of the tubes. The liquor absorbs heat from steam condensing on the outside of the tubes, and water within the liquor is vaporized. The vapor and liquor are then separated at the bottom of the tubes (Figure 8.10).

Vertical tube falling film evaporators are typically used on higher viscosity liquors and for concentrating heat-sensitive solutions that require low residence times.

For applications in which the solubility of silica, sulfate, or other scaling constituents may be exceeded, the vertical tube evaporator may be operated with an excess concentration of calcium sulfate seed crystals in the liquor. The seed crystals will serve as nucleation sites for the precipitation of scale. As the solution becomes supersaturated because of pressure, temperature,

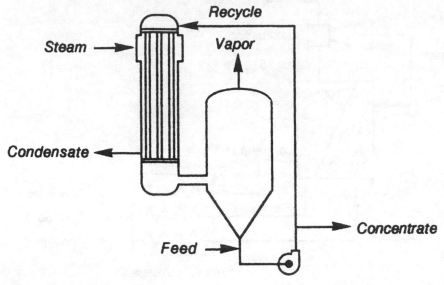

Figure 8.10   Falling film evaporator.

and continuous water evaporation, scale deposits on the existing seed crystals and not on the heat-transfer surfaces. With high recirculation rates and evaporation rates greater than 98% of the feed flow, the amount of concentrate blowdown is low, and the proper concentration of seed crystals can be maintained.

Horizontal Tube Spray Film. Recirculating liquor is heated and sprayed over the outside of a horizontal tube bundle carrying low-pressure steam and condensed water vapor inside the tubes. Vapor from the evaporator chamber can be used as steam in a subsequent effect or mechanically compressed and reused as the heating medium for the stage in which it was generated (Figure 8.11).

Scale forming on the outside of the tubes can be periodically removed through chemical cleaning. Horizontal tube designs can be applied indoors or in locations with low headroom requirements.

Forced Circulation. Recirculating liquor is pumped through a heat exchanger under pressure to prevent boiling and subsequent scale formation in the tubes. The liquor then enters a separator chamber operating at a slightly lower pressure or partial vacuum, causing flash evaporation of water and formation of insoluble crystals in the liquor (Figure 8.12).

Forced circulation evaporators, or crystallizers, are often used for applications requiring high solids concentrations or crystallizing, or in applications having large amounts of suspended solids.

Energy costs typically are higher for forced-circulation units than for other types of evaporators because of the high recirculation rates necessary.

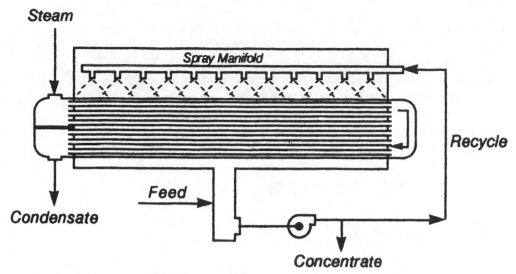

Figure 8.11   Spray film evaporator.

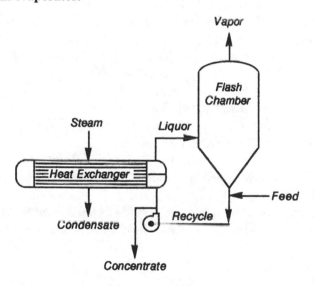

Figure 8.12   Forced-circulation evaporator.

Combined Systems. It is sometimes possible to combine different types of evaporators or evaporators with other treatment processes to reduce capital equipment and operating costs or meet specific treatment objectives.

One fairly common arrangement uses a falling-film evaporator followed by a forced-circulation crystallizer. In this scheme, an evaporator initially concentrates the waste to 20 to 30% solids, and a crystallizer further concentrates it to a solid form. Energy costs may be reduced by using vent steam from the evaporator to operate the crystallizer.

"Hybrid" systems are being considered in an increasing number of zero-liquid-discharge applications. A hybrid system consists of an evaporator or evaporator/crystallizer preceded by a reverse osmosis or electrodialysis pre-concentration step. In this arrangement, concentrate, or reject, from the pre-concentrator becomes the feed for the evaporator.

Although a hybrid system introduces an additional technology that may add to the complexity of the system, it should significantly reduce the size of the evaporator unit and result in a reduction of the system's overall energy requirement. It should be noted that some wastewaters, especially those with high scaling tendencies, may not be candidates for hybrid systems.

REFERENCES

Metcalf & Eddy, Inc. (1991) *Wastewater Engineering: Treatment, Disposal, Reuse.* McGraw-Hill, Inc., New York, N.Y.

Patterson, J.W. (1975) *Wastewater Treatment Technology.* Ann Arbor Science Publishers, Inc., Ann Arbor, Mich.

Sittig, M. (1978) *Electroplating and Related Metal Finishing Pollutant and Toxic Materials Control.* Noyes Data Corp., Park Ridge, N.J.

U.S. Environmental Protection Agency (1979) *Draft Development Document for Effluent Limitations Guidelines and Standards for the Porcelain Enameling Point Source Category.* Effluent Guidelines Div., Office of Water and Haz. Mater., Washington, D.C.

U.S. Environmental Protection Agency (1980) *Control and Treatment Technology for the Metal Finishing Industry—Sulfide Precipitation.* Technol. Transfer, Washington, D.C.

U.S. Environmental Protection Agency (1985) *Environmental Regulations and Technology—The Electroplating Industry.* EPA/625/10-85/001, Washington, D.C.

Chapter 9
Treatment of Organic Constituents

Many industries produce wastewater that contains significant concentrations of organic constituents, measured as biochemical oxygen demand (BOD) and/or chemical oxygen demand (COD). These organic materials may be present as suspended, floating, and/or dissolved constituents. Table 9.1 presents a list of example industries and general indications of the composition of their wastewaters. The wastewaters of many of these industries (for example, food processing and beverage) contain organic compounds that are treated effectively by a municipal wastewater treatment plant (WWTP) and cause no significant problems. These organics are classified as "compatible" substances. The wastewaters of some industries, however, contain organic compounds that are included in the priority pollutant list (for example, the textile, petroleum refining, pharmaceutical, organic chemicals, and coke-manufacturing industries all contain polynuclear aromatic compounds and

Table 9.1 Industrial wastewater characteristics (U.S. EPA, 1976).

Industry	Flow	BOD	TSS	COD	pH	Grit	Nitrogen	Phosphorus	Heavy metals	Chlorine demand
Meat products	Intermittent	High-extremely high	High	High-extremely high	Neutral	Absent	Present	Present	Absent	High
Dairy products										
Milk handling and milk products	Intermittent	Average-high	Low-average	Average-high	Acid-alkaline	Present	Adequate	Present	Absent	High
Natural and cottage cheese products	Intermittent	Extremely high	Average-extremely high	Extremely high	Acid-alkaline	Present	Deficient	Present	Absent	High
Beverages										
Malt beverages and distilled spirits	Intermittent to continuous	High	Low-high	High	Acid-neutral Acid-alkaline	Present	Deficient	Deficient	Absent	No data
Wine and brandy	Intermittent	High-extremely high	Low-high	High-extremely high	Alkaline	Present	Deficient	Deficient	Absent	No data
Soft drinks bottling	Intermittent	Average-high	Low-high	Average-high		Present	Deficient	Present	Absent	No data
Textiles										
Wool	Intermittent to continuous	High	High	High	Alkaline	Present	Deficient	Present	Present	High
Cotton and synthetics	Continuous	Average-high	Low-average	Average-high	Alkaline	Absent	Deficient	Present	Present	High
Leather tanning and finishing										
Chrome tanning	Intermittent	Extremely high	Extremely high	Extremely high	Acid-alkaline	Present	Adequate	Deficient	Present	High
Vegetable tanning	Intermittent	Extremely high	Extremely high	Extremely high	Acid-alkaline	Present	Adequate	Deficient	Present	High
Petroleum refining	Continuous	Average	Low	High	Acid-alkaline	Present	Adequate	Deficient	Absent	High
Metal finishing	Continuous-variable	Low	Average-high	Low	Acidic	Present	Present	Present	High	High

Table 9.1 Industrial wastewater characteristics (continued).

Industry	Flow	BOD	TSS	COD	pH	Grit	Nitrogen	Phos-phorus	Heavy metals	Chlorine demand
Fruit and vegetables	Intermittent	Average-extremely high	Average-extremely high	—	Acid-alkaline	Present	Deficient	Deficient	Absent	Average-high
Paper and allied products										
Mechanical pulping	Continuous	Extremely high	High	High	Neutral	Present	Deficient	Deficient	Absent	High
Chemical pulping (unbleached)	Continuous	Average-extremely high	Low-high	Low-high	Acid-alkaline	Present	Deficient	Deficient	Present	High
Chemical pulping (bleached)	—	Average-extremely high	Low-high	High	Acid-alkaline	Present	Deficient	Deficient	Present	High
Pharmaceutical	Continuous-variable	High	Low-High	High	Acid-alkaline	—	—	—	—	—
Organic Chemicals										
Plastics and resins	Continuous-variable	Average-high	Low-high	Average-high	Acid-alkaline	—	—	—	—	Average-high
Coke	Continuous-variable	High	Average	High	Neutral-alkaline	—	Present	—	Present	Average-high

phenolic and halogenated hydrocarbons). Some of these wastewaters may contain organic compounds that are "incompatible" with the processing at a municipal wastewater treatment facility. Examples of these potentially incompatible compounds are

- Biologically recalcitrant organic dyes that pass through the treatment facility;
- High-molecular-weight polynuclear aromatic hydrocarbons (5- and 6-ring structures)—these are slowly degraded and tend to accumulate in the biomass, thus increasing the problem of sludge disposal; and
- Volatile organic compounds—these could be discharged to the atmosphere, causing air emissions problems.

This chapter deals primarily with the treatment of wastewater constituents that are compatible with treatment provided by a WWTP, although some of the treatment technologies discussed can provide pretreatment for "incompatible" constituents if appropriately designed.

*S*ELECTION OF PRETREATMENT TECHNOLOGY

Selection of the most effective pretreatment system depends on the following:

- Wastewater characteristics
    - Flow: batch or continuous, frequency, volume, and rate.
    - Chemical composition: high strength versus low, solids to dissolve, and biodegradable or nonbiodegradable.
- Pretreatment limits
    - Categorical standards.
    - Pretreatment ordinance limits.
- Site-specific constraints
    - Available land for treatment facility.
    - Proximity to residential community.
    - Level of technical skill available for operating treatment facility.
    - Anticipated need for future expansion.
- Capital and operating costs.

For the list of manufacturing operations in Table 9.1, Table 9.2 presents alternative pretreatment technologies. The information in Table 9.2 shows that treatment of a given wastewater can be accomplished by a train of treatment processes, for example, equalization followed by screening, gravity separation, and biotreatment. Also, the biotreatment stage may be replaced by, or

Table 9.2 Pretreatment alternatives for compatible organics in selected industrial wastes.

Industry	Pretreatment alternatives
Food processing	
Dairies	Equalization, biological treatment (aerobic/anaerobic), removal of whey
Meat, poultry, and fish	Screening, gravity separation, flotation, coagulation/precipitation, biological treatment (aerobic/anaerobic)
Fruit and vegetable canneries	Screening, equalization, gravity separation, neutralization, biological treatment, coagulation/precipitation
Breweries and distilleries	Screening, centrifugation, biological treatment (aerobic/anaerobic)
Pharmaceuticals	Equalization, neutralization, coagulation, solvent extraction, gravity separation, biological treatment
Organic chemicals	Gravity separation, flotation, equalization, neutralization, coagulation, oxidation, biological treatment, adsorption
Petroleum refineries	Gravity separation, flotation, equalization, coagulation, chemical oxidation, biological treatment, adsorption
Pulp and paper mills	Screening, gravity separation, biological treatment, chemical oxidation
Plastics and resins	Gravity separation, flotation, coagulation, chemical oxidation, solvent extraction, adsorption, biological treatment
Explosives	Flotation, chemical precipitation, biological treatment
Rubber	Chemical oxidation, biological treatment
Textiles	Equalization, neutralization, coagulation, adsorption, biological treatment, ultrafiltration
Leather tanning and finishing	Screening, gravity separation, flotation, coagulation, neutralization, biological treatment (aerobic/anaerobic)
Coke and gas	Equalization, flotation, adsorption, biological and chemical oxidation, solvent extraction

complemented with, chemical oxidation or carbon adsorption. Where alternative technologies are available, costs and site-specific constraints become key factors in technology selection. Generally, biological treatment has tended to be more cost effective for organic constituents than chemical or physical treatment technologies.

TECHNOLOGIES

The pretreatment schemes shown in Table 9.2 comprise technologies that can be categorized as biological, chemical, or physical.

BIOLOGICAL TREATMENT. Biological treatment depends on the ability of microorganisms to use (degrade) wastewater constituents in their metabolic processes. Typically, wastewater constituents amenable to biological treatment are organic compounds and certain inorganic constituents (for example, ammonia, cyanide, sulfide, sulfate, and thiocyanate). Biological treatment of organics can be expressed as:

$$\text{Organic matter} + \text{Electron acceptor} + \text{Micronutrients} + \text{Biomass} \rightarrow \text{Carbon dioxide} + \text{Water} + \text{Energy}$$

The energy generated can be used to synthesize new cell mass.

The biological process varies primarily with the electron acceptor used by the biomass in respiration. Examples of biological processes are

- Aerobic process using oxygen;
- Anaerobic (absence of oxygen) process using sulfate, phosphate, or other organics;
- Anoxic process using nitrate; and
- Facultative process using oxygen and nonaerobic electron acceptors.

Microbial metabolic processes include cell-mass synthesis, which requires the availability of macronutrients nitrogen and phosphorus, the primary constituents of protein. Minimum nutrient requirements have been established as ratios of BOD:nitrogen:phosphorus of 100:5:1 (Sawyer, 1956). This ratio varies, however, for certain industrial wastewaters.

Processing conditions such as pH and temperature can affect biological treatment. Most biological treatment systems function effectively within a pH range of 6.0 to 8.5; the anaerobic (methanogenic) process has a narrower operating range of 6.5 to 7.5 (Pohland, 1967). Biological treatment is more dependent on temperature, however, with activity in three temperature zones: a psychrophilic range of 0 to 20°C, a mesophilic range of 20 to 45°C, and a thermophilic range of 45°C and higher. These reaction rates can be approximated by the following mathematical expression:

$$k_t = k_{20}\,\theta^{(t-20)}$$ (9.1)

Where

k_t	=	reaction rate at t°C;
k_{20}	=	reaction rate at 20°C;
t	=	temperature, °C; and
θ	=	activity coefficient with values of 1.0 to 1.03 for activated sludge, 1.02 to 1.04 for trickling filters, and 1.06 to 1.09 for aerated lagoons (Eckenfelder, 1970).

Heavy metal ions may inhibit the activity in a biological treatment process. The extent of the inhibition depends on the concentration of metal ion in solution and the opportunity or lack of opportunity for acclimation. Table 1.8 presents reported threshold concentrations of various heavy metals in the activated sludge process. Heavy metal concentrations greater than the threshold concentrations will inhibit the biological treatment process. See Chapter 8 for heavy metals removal information.

The priority pollutant list includes 114 organic compounds (see Tables 1.1 to 1.6). Biological acclimation/degradation tests of 97 of these compounds showed that only eight compounds were biodegraded less than 5% after the third subculture (Tabak et al., 1981). Table 9.3 shows the results of the biotreatability tests. Tables 9.4 and 9.5 present reported results of organic priority pollutant removal in biological treatment systems. While these data show that a large number of the organic priority pollutants are amenable to biological treatment, it is important to note that removal mechanisms, sorption and volatilization, are significant and can occur with many of these compounds to varying degrees.

Many surfactants today are biodegradable. Treatment in an activated sludge basin may necessitate some simultaneous defoamer use. Recent research has shown that the popular linear alkylbenzene sulfonate (LAS) has a 95 to 99% removal (biodegradation) rate in average secondary wastewater treatment processes (Weber, 1972). Oxidation sometimes is an alternative step.

Process Configuration. The microorganisms used in biological treatment have the ability to grow in an aqueous suspension and/or while attached to a support surface. These capabilities have given rise to suspended growth reactors (for example, the activated sludge process) and fixed film reactors (for

Table 9.3   Bioacclimation/degradation of priority pollutants (Patterson, 1985).*

	Biodegradation, %	
	Original culture	Third subculture
Benzene	49	100
Carbon tetrachloride	87	100
Chlorobenzene	89	100
1,2,4-Trichlorobenzene	54	24
Hexachlorobenzene	56	5
1,2-Dichloroethane	26	63
1,1,1-Trichlorobenzene	29	83
1,1-Dichloroethane	50	91
1,1,2-Trichloroethane	6	44
1,1,2,2,-Tetrachloroethane	0	29
2-Chloroethyl vinyl ether	76	100
Para-chloro-meta-cresol	78	100
Chloroform	49	100
2-Chlorophenol	86	100
1,2-Dichlorobenzene	45	29
2,3-Dichlorobenzene	59	35
1,4-Dichlorobenzene	55	16
1,1-Dichloroethylene	78	100
1,2-*trans*-Dichloroethylene	67	95
1,2-Dichloropropane	42	89
1,3-Dichloropropylene	55	85
2,4-Dinitrotoluene	77	27
2,6-Dinitrotoluene	82	29
1,2-Diphenylhydrazine	80	77
Fluoranthene	0	100
4-Chlorophenyl phenyl ether	0	1
4-Bromophenyl phenyl ether	0	0
Bis-(2-chloroisopropyl) ether	85	100
Bis-(2-chloroethoxy) methane	0	0
Bromoform	11	48
Dichlorobromomethane	35	59
Trichlorofluoromethane	59	73
Chlorodibromomethane	25	39
2,4-Dinitrophenol	60	100
4,6-Dinitro-*o*-cresol	52	51
Nitrosodimethylamine	71	57
N-Nitrosodiphenylamine	87	100
N-Nitroso-di-*n*-propylamine	27	50
Pentachlorophenol	19	100
Bis-(2-ethyl hexyl)phthalate	0	95

Table 9.3   Bioacclimation/degradation of priority pollutants*
(continued).

	Biodegradation, %	
	Original culture	Third subculture
Di-*n*-octyl phthalate	0	92
Benzo(a)anthracene	16	0
Chrysene	0	38
Anthracene	43	92
Fluorine	82	77
Pyrene	71	100
Tetrachloroethylene	45	87
Trichloroethylene	64	87
PCB-1242	37	66
PCB-1254	11	0
PCB-1248	0	0
PCB-1260	0	0
PCB-1016	44	48

* All priority pollutants not listed in this table, except for those identified below, exhibited 90% or greater degradation by the original inoculum. Priority pollutants not tested were

Benzidine,	Dichlorodifluoromethane,
Benzo(a)pyrene,	Dioxin,
3-4-Benzofluoranthene,	Endrin aldehyde,
Benzo(g,h)perylene,	Indeno(1,2,3-cd)pyrene,
Benzo(k)fluoranthene,	Methyl bromide,
Bis-(2-chloromethyl)ether,	Methyl chloride,
Chloroethane,	Toxaphene, and
Dibenzo(a,h)anthracene,	Vinyl chloride.
3,3-Dichlorobenzidine,	

All pesticides tested exhibited zero degradation through the third subculture.

example, trickling filters and rotating biological discs). In either type of reactor, the pretreatment facility could include the following processing stages:

- Primary treatment to remove excess oil and grease and/or gross suspended solids.
- Biocontactor (aeration basin, trickling filter, rotating biodiscs, or digester), where microorganisms degrade/transform the organic matter.

Table 9.4 Removal of organic priority pollutants in biological treatment systems (Patterson, 1985).

	Type of treatment[a]	Influent level, μg/L	Effluent level, μg/L	Removal, %
Acenaphthene	AS–M	0.9	—	95
	AS–l(10)	15	—	84+
	AS–l(5)	84	5	94
	AL–l(1)	4.0	—	0
Acrolein	AS–l[b]	62 000	50	99+
	AS–l[b]	31 000	30	99+
Acrylonitrile	AS–l(5)	10 300	65	99+
	AS–l[b]	152 000	50	99+
	AS–l[b]	31 000	30	99+
Benzene	AS–M	8.8		66
	AS–M	7.7		92
	AS	—		90–100
	AL	6–53		(−17)–96
	AS–M	6.8	3.5	49
	PACT™	160	—	99+
	PACT™	105	0.9	99
	AS–l(9)	10 250	—	60+
	AS–l(5)	581	6	99+
	AS–l[b]	153 000	40	99+
	AS–l[b]	39 000	<50	99+
	AL–l(2)	46.9	—	84+
Benzidine	AS–l	4	—	0
	AL–l	11.9	—	41
Carbon tetrachloride	AS–M	1.0	nd[c]	100
	PACT™	94	1.4	95
	PACT™	95	—	94
	AS–l(2)	250	—	98+
	AS–l(5)	51	nd	100
	AS–l[b]	7 000	7	99+
Chlorobenzene	PACT™	1 720	30	98
	PACT™	1 900	—	99+
	AS–M	177	—	99+
	AS–l(6)	15.2	—	67+
	AS–l(5)	20	16	20
1,2,4–Trichlorobenzene	PACT™	523	169	66
	AS–M	0.6	—	83
	As–l(ll)	285	—	67+
	AS–l(5)	234	39	83
Hexachlorobenzene	AS–l(4)	0.75	—	47+
	AL–l	10	—	0

Table 9.4 Removal of organic priority pollutants in biological treatment systems (continued).

	Type of treatment[a]	Influent level, µg/L	Effluent level, µg/L	Removal, %
1,2–Dichloroethane	AS–M	1.5	—	65
	AS–M	8.4	—	85
	AS–l(5)	524	9	98
	AS–l[b]	63 000	1 000	98
	AS–l[b]	258 000	3 700	99
1,1,1–Trichloroethane	PACT™	13	0.6	89
	PACT™	18	—	99+
	AS–M	88.4	—	70
	AS–M	1 791	—	98
	AS–M	14	5.2	83
	AS–l(6)	9.2	—	74+
	AL–1	550	—	96
	AS–l(5)	7	5	29
	AS–l[b]	60 000	94	99+
	RBC–1	42	22	48
1,1–Dichloroethane	AS–M	1.7	—	6
	AS–M	6.7	—	92
	AS–l(2)	7.1	—	9+
	AS–l(5)	9	5	44
1,1,2–Trichloroethane	AS–M	5.2	nd	100
	AS–l	11	—	9+
	AS–l(5)	12	5	58
1,1,2,2–Tetrachloroethane	AS–l(2)	12.8	—	22+
	AS–l(5)	17	5	71
	AS–l[b]	52 000	2 000	96
	AS–l[b]	201 000	13 000	94
Chloroethane	PACT™	280	12	94
	AS–1(5)	12	5	58
Bis(chloromethyl)ether	AS–l	59	—	83
Bis(2–chloroethyl)ether	AS–M	0.12	—	02
	AS–l	19	—	47+
Naphthalene	AS[b]	2 000	—	100
	AS–l	2	—	50
	TF–l	2	—	0
	AL–l	19	—	> 47
2,4,6–Trichlorophenol	AS–l(10)	703	—	36
	AS–l(5)	100	53	47
	AL–l	1 000	—	99

Table 9.4 Removal of organic priority pollutants in biological treatment systems (continued).

	Type of treatment[a]	Influent level, μg/L	Effluent level, μg/L	Removal, %
Chloroform	PACT™	201	21	81
	AS–M	37.5	16	57
	AS–M	4.2	—	5
	AS–M	17.1	—	98
	AS–M	37.5	—	49
	AS–1(16)	33	—	61+
	AS–1(5)	348	13	96
	AS–1[b]	14 000	105	99+
	TF–1	19	—	0
	AL–1(3)	531	—	36+
	AL–1	425–2 645	—	(–1–86)
	RBC–1	8	4	50
Chlorodibromomethane	PACT™	81	—	99+
	AS–1(5)	6	nd	100
Isophorone	AS–1(2)	10	—	> 0
	AS–1(5)	660	nd	100
	AL–1	3.0	—	33
Naphthalene	AS–M	3.2	3.2	0
	AS–M	1.3	—	35
	AS–M	10.9	—	62
	AS–1(26)	50	—	64+
	AS–1(5)	802	6	99
	AS–1[b]	2 300	< 50	97+
	TF–1	55	—	0
	AL–1(2)	14.1	—	29+
Nitrobenzene	PACT™	454	2	99+
	AS–1(5)	3 000	91	97
	AS–1[b]	37 500	24 800	34
	AS–1[b]	100 000	2 200	98
	AL–1	10	—	0
2–Nitrophenol	AS–1	40	—	99+
	AS–1(5)	40	9	78
4–Nitrophenol	PACT™	56	—	99+
	PACT™	1 020	10	97
	AS–1	90	—	99+
	AL–1	13	—	23
1,4–Dinitrophenol	PACT™	161	5	98
	AS–1(5)	673	59	91
2,6–Dinitro–o–cresol	PACT™	11	—	99

Table 9.4 Removal of organic priority pollutants in biological treatment systems (continued).

	Type of treatment[a]	Influent level, μg/L	Effluent level, μg/L	Removal, %
Nitrosodimethylamine	AS[b]	2 000	—	95+
N–Nitroso–diphenylamine	AS–1	5.3	—	84+
	AL–1	3	—	67
N–Nitrosodi–n–propylamine	AS–M	0.5	—	88
	AS–Ml	6.7	—	99
	AS–1(2)	11	—	0
Pentachlorophenol	AS–1(15)	5 333	—	70+
	AS–1(5)	216	115	47
	AS–1[b]	3 700	83	98
Phenol	TF	105		91
	AS	105		> 99
	PACT™	440		96
	AS–1(30)	335		> 77
	TF–1	37		0
Bis(2–ethylhexyl)phthalate	PACT™	2.4	—	67
	AS–M	25.5	12.0	53
	AS–M	1.3	—	31
	AS–M	50	—	78
	AS–1(38)	102	—	37
	TF–1	35	—	83
	AL–1(5)	37	70+	1
	TL–1(2)	26	—	58+
	AS–1(5)	24	28	(−16)
	AS–1[b]	350	50	86
Butyl benzyl phthalate	AS–M	1.7	3.2	(−88)
	AS–M	18	—	91
	AS–1	11	—	0
	AS–1(5)	86	6	93
	AL–1	6	—	0
Di–n–butyl phthalate	AS–M	4.3	4.3	0
	AS–M	5.6	—	68
	AS–M	13	—	62
	AS–1(9)	23	—	> 60
	AS–1(5)	86	6	93
	TF–1	15	—	25
	AL–1	1	—	0
Di–n–octyl phthalate	AS–1	5 000	—	0
	AS–1(15)	28	5	82

Table 9.4   Removal of organic priority pollutants in biological treatment systems (continued).

	Type of treatment[a]	Influent level, μg/L	Effluent level, μg/L	Removal, %
Diethyl phthalate	AS–M	6.6	—	93
	AS–M	1.4	—	79
	AS–l(17)	15	—	56
	AS–l(5)	134	9	93
	TF–1	140	—	0
	AL–l	4	—	0
Dimethyl phthalate	AS–l(9)	60	—	60
	AS–l(5)	46	5	89
	AL–l	8	—	25
Benzo(a)pyrene	AL–l	3	—	33
	AS–l(15)	13	18	(−38)
Acenaphthylene	AS–M	0.2	—	80
	AS–l(5)	65	5	92
	AL–l	5	—	0
Anthracene	AS–l(7)	8.5	—	> 60
Fluorene	AS–M	1.7	—	94
	AS–l(2)	2	—	99+
	AS–l(5)	56	5	91
Phenanthrene	AS–M	3.2	—	81
	AL–l	3	—	0
	AS–l[b]	560	< 10	98+
2–Chlorophenol	PACT™	11	1.6	95
	AS–l(2)	10	—	46
	AS–l(5)	53	35	34
Dichlorobenzene	PACT™	720	—	35
1,2–Dichlorobenzene	PACT™	259	120	44
	AS–M	1.8	—	28
	AS–M	4.8	—	28
	AS–M	43.9	—	97
	AS–l(12)	28.8	—	74+
	AS–l(5)	331	28	92
	AS–l[b]	10 000	50	99+
1,3–Dichlorobenzene	AS–M	3.1	—	86
	AS–l[b]	30 000	2.5	99+
	AS–l[b]	83 000	50	99+
1,4–Dichlorobenzene	AS–l(8)	30	—	82+
	AL–1	53	—	81+
1,1–Dichloroethylene	AS–M	43.2	—	97
	AS–M	3.2	5.4	(−69)

Table 9.4   Removal of organic priority pollutants in biological treatment systems (continued).

	Type of treatment[a]	Influent level, μg/L	Effluent level, μg/L	Removal, %
1,2-*trans*-Dichloroethylene	AS–M	3.0	1.3	57
	AS–1(5)	42	5	88
2,4–Dichlorophenol	AS–1(2)	13.3	—	25+
	AS–1(5)	347	24	93
	AS–1[b]	24 100	5 500	77
	AS–1[b]	75 000	3 600	95
1,2–Dichloropropane	AS–1(2)	16	—	67+
	AS–1(5)	138	7	95
	AS–1[b]	48 000	170	99+
	AS–1[b]	182 000	1 810	99
1,3–Dichloropropene	AS–1(5)	148	6	96
2,4–Dimethylphenol	AS–1(3)	39.7	—	32+
	AS–1(5)	270	5	98
2,4–Dinitrotoluene	PACT™	2 000	—	75
	PACT™	1 900	243	65
	AS–1[b]	32 000	250	99+
	AL–1	3	—	0
2,6–Dinitrotoluene	PACT™	1 640	575	64
	PACT™	1 900	—	76
	AS–1	390	—	0
	AL–1	12	—	83+
1,1–Diphenylhydrazine	AS[b]	2 400	—	100
	AS–1	340	—	0
	AL–1	14	—	0
Ethylbenzene	PACT™	41	1.7	94
	PACT™	29	—	78
	AS–M	148	—	99
	AS–M	25.5	5.1	80
	AS–1(24)	882	—	83
	AS–1(5)	283	5	98
	AS–1[b]	19 300	< 10	99+
	AL–1(3)	45	—	78+
Fluoranthene	AS–M	0.6		92
	AS–1	2	—	0
	AL–1	2	—	0
	AS–1(5)	17	6	65
4–Bromophenyl phenyl ether	AS–1	360	—	95
Bis(2–chloroisopropyl)ether	AS–M	0.16	—	62
	AL–1	2	—	0

Table 9.4   Removal of organic priority pollutants in biological treatment systems (continued).

	Type of treatment[a]	Influent level, μg/L	Effluent level, μg/L	Removal, %
Bis(2–chloroethoxy)methane	AL–l	25	—	60+
Methylene chloride	AS–M	7.1	7.8	(−10)
	AS–M	9.0	—	49
	AS–M	293	—	0
	AS–M	647	—	72
	AS–l(5)	144	—	34
	AS–l(5)	17	9	47
	AS–l[b]	56 000	50	99+
	AS–l[b]	180 000	51	99+
	TF–l	1	—	0
	AL–l(3)	1 114	—	65
Methyl chloride	PACT™	1 770	nd	100
	AL–l	56	—	91+
Bromomethane	AS–l(5)	1 250	nd	100
Bromoform	PACT™	910	—	89
	AS–l	3	—	0
Dichlorobromomethane	PACT™	54	—	99+
	AS–l	5.8	—	0
	AS–l(5)	20	5	75
Trichlorofluoromethane	PACT™	155	3.0	95
	PACT™	920	—	99
	AS–l(5)	556	—	19
	TF–l	48	—	79+
Dibenzo(a,h)anthracene	AS–l(5)	15	75	(−400)
Pyrene	AS–l(5)	17	6	65
	AS–l(5)	2.3	0	16
	AL–l	3	—	67
Tetrachloroethylene	PACT™	24	1.7	93
	PACT™	62	—	88
	AS–M	6.4	—	0
	AS–M	49	—	80
	AS–M	560	—	99+
	AS–M	41.5	0.7	84
	AS–l(11)	25.6	—	75+
	RBC–l	23	16	30
	AL–l	25	—	60+
Toluene	PACT™	519	1.7	99+
	PACT™	680	—	99+
	AS–M	70	—	99+

Table 9.4 Removal of organic priority pollutants in biological treatment systems (continued).

	Type of treatment[a]	Influent level, μg/L	Effluent level, μg/L	Removal, %
Toluene (continued)	AS–M	112	—	99
	AS–M	15.5	6.5	58
	AS–I(31)	119	—	52
	AS–I(5)	4 500	7	99+
	AS–I[b]	16 800	< 10	99+
Trichloroethylene	AS–M	20	0.3	69
	AL	10–120	—	(−23)–73
	PACT™	60	—	90
	AS–I(12)	31.3		68
	TF–I	1	—	0
	RBC–I	210	31	85
Vinyl chloride	AS–I(5)	6	nd	100
Aldrin	AS + MMF[b]	—	—	100
Dieldrin	AS + MMF[b]	—	—	72
DDT	AS + MMF[b]	—	—	100
DDE	AS + MMF[b]	—	—	39
DDD	AS + MMF[b]	—	—	100
	AS–M	0.3	—	37
	AS–M	0.17	—	17
Heptachlor	AS–I	6.3	—	76
Landane	AS + MMF[b]	—	—	0
PCB	TF–1	524	6.5	99
	TF–M	30–470	10–50	33–89

[a] AS = activated sludge; M = municipal facility; I = industrial facility; AL = aerated lagoon; TF = trickling filter; MMF = mixed-media filtration; PACT™ = activated sludge with powdered activated carbon added; TL = tertiary polishing lagoon; RBC = rotating biological contactor. Numbers in parentheses represent number of treatment plants for which data were averaged.

[b] Pilot-scale data.

[c] Nondetectable.

Table 9.5 Organic priority pollutant removal in activated sludge systems (Patterson, 1985).

	Influent, µg/L	Percent removal	
		Primary	Total
Acenaphthene	39.8	(−35)	> 97
Benzene	33.7	19	99
	18	7	80
Carbon tetrachloride	38.2	11	99
	160	13	> 99
Chlorobenzene	171	13	> 99
1,2,4-Trichlorobenzene	17	12	82
1,1,1-Trichloroethane	43.3	15	97
	20	13	82
1,1-Dichloroethane	60.5	37	> 99
1,1,2-Trichloroethane	65.7	27	69
Chloroform	19	21	89
	73	1	56
	47.6	20	97
1,3-Dichlorobenzene	3	14	40
1,4-Dichlorobenzene	6	0	88
1,2-*trans*-Dichloroethylene	1	0	100
1,2-Dichloropropane	54.5	19	97
2,4-Dimethylphenol	95.7	36	99
Ethylbenzene	47.8	29	99
	23	0	89
Fluoranthene	30.6	(−30)	> 94
Methylene chloride	106	39	99
	89	16	55
Dichlorobromomethane	28.6	24	98
Chlorodibromomethane	37.2	37	73
Naphthalene	12	0	92
2,4-Dichlorophenol	6	2	46
Pentachlorophenol	7.6	(−71)	> 47
Bis(2-ethylhexyl)phthalate	51.7	(−1)	78
	12	0	77
Di-*n*-butylphthalate	43.8	(−24)	94
	6	40	44
Di-*n*-octylphthalate	28.2	(−22)	> 83
Diethylphthalate	46.4	(−24)	> 97
	4	0	36
Dimethylphthalate	47.3	> 21	> 98
Benzo(a)anthracene	23.8	(−5)	> 97
	31.4	81	> 99

Table 9.5   Organic priority pollutant removal in activated sludge systems (continued).

	Influent, µg/L	Percent removal	
		Primary	Total
Benzo(a)pyrene	34.5	96	> 99
Benzo(k)fluoranthene	8.1	73	> 99
Chrysene	38.9	6	> 97
Anthracene	34.8	3	> 97
Benz(ghi)perylene	8.7	86	> 99
Fluorene	37.9	(−36)	> 98
Phenanthrene	40.4	(−10)	> 98
Indeno(1,2,3-cd)pyrene	15.0	80	> 99
Pyrene	11.8	81	> 99
	30.4	(−29)	> 93
Tetrachloroethylene	52	25	91
	7	(−29)	57
Toluene	321	10	87
	195	24	95
Heptachlor	31.7	> 10	> 93
Lindane	45.5	8	> 43
PCB-1254	33.5	(−240)	> 91
Toxaphene	47.4	(−85)	> 94

- Clarifier, where the biomass (sludge) is separated from the treated wastewater. A portion of the separated sludge is returned to the bio-contactor for activated sludge, and sludge is removed from the system for further processing and disposal.

Biological treatment of wastewater can be accomplished through a variety of process configurations. Each configuration has advantages and disadvantages. Such processes may be similar for pretreatment and direct discharge, with degree of treatment, economics, and political factors influencing the choice.

Figure 9.1 presents schematic flow diagrams of typical biological treatment systems. In some cases, biocontactor and clarifier operations have been combined to operate from a single processing stage (for example, sequencing batch reactors and lagoons/ponds). With a sequencing batch reactor, wastewater is charged in batches, and each processing stage is allowed a predetermined time period for completion. Typically, these systems consist of at least two reactors. They are effective for small or intermittent waste-water-generation situations. Design of lagoons and ponds must involve

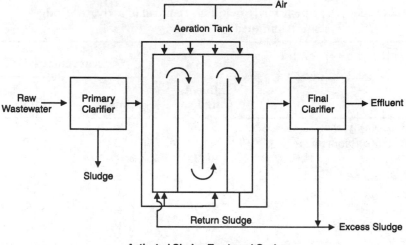

Activated Sludge Treatment System

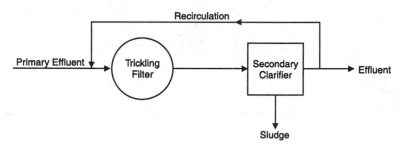

Trickling Filter Treatment System

Figure 9.1   Schematic flow diagrams of typical biological treatment systems.

groundwater and soil protection criteria for liner selection and system optimization.

Selection of an appropriate aerobic or anaerobic treatment system involving fixed film or suspended growth can be guided by the criteria in Tables 9.6 and 9.7.

Secondary Emissions. Secondary emissions from wastewater treatment facilities are governed by various agencies and regulations such as the Clean Air Act Amendments of 1990. For example, U.S. EPA promulgated final standards under National Emission Standards for Hazardous Air Pollutants for benzene and vinyl chloride emissions from wastewater treatment facilities. Increased regulation of secondary emissions should be anticipated.

Table 9.6 Criteria for selecting an aerobic or anaerobic pretreatment process.

Parameter	Aerobic treatment	Anaerobic treatment
Influent characteristics	< 5% TDS < 2 000 mg/L BOD No alkalinity requirement 5 lb NH_3–N/100 lb BOD[a] 1 lb PO_4–P/100 lb BOD Sulfur compounds compatible 5°C to 30°C	< 5% TDS > 2 000 mg/L BOD > 1 500 mg/L $CaCO_3$ 2 lb NH_3–N/100 lb BOD 0.4 lb PO_4–P/100 lb BOD Sulfur compounds not compatible 29°C to 38°C or 49°C to 57°C
Oxygen requirement	Waste dependent	None
Reactor conditions	6.5 to 8.5 pH > 2.0 mg/L DO	6.5 to 7.5 pH No oxygen requirement
Sludge production	0.6 to 1.1 lb TSS/lb BOD removed	0.2 to 0.4 lb TSS/lb BOD_R
Methane production	None	Approximately 4 cu ft/lb BOD_R[b]
Typical biochemical oxygen demand removal	85 to 99.9%	80 to 85%
Advantages	Greater process variability tolerance, quicker process recovery, less space, and no alkalinity requirement	Less energy and nutrient requirements and sludge production. Recoverable energy.
Disadvantages	Greater foaming potential, more uncollected off gas, greater sludge bulking potential, more energy intensive, and poorer sludge dewaterability.	Higher risk of corrosion problems and septic odors, longer acclimation time.

[a] lb/lb × 1 000 = g/kg.
[b] cu ft/lb (6.243×10^{-2}) = m^3/kg.

Table 9.7   Criteria for choosing between fixed film and suspended growth.

Parameter	Fixed film treatment	Suspended growth treatment
Advantages	Smaller land requirements. Less offgas for volume. Lowering sludge bulking potential for aerobic systems. Improved sludge dewaterability. Simpler to operate for aerobic systems. Less energy intensive.	Can tolerate greater variations in influent characteristics. Can tolerate high influent TSS concentrations. Simpler to operate for anaerobic systems. Quicker recovery times. Greater process flexibility.
Disadvantages	Typically requires influent pumping. Presence of high oil and grease concentrations (> 50 mg/L) fouls media. Presence of high TSS concentrations (> 200 mg/L) may foul distribution mechanism and media filter. Fly annoyance for open aerobic systems. Greater effect from temperature changes.	Larger land requirements. Greater uncollected offgas volumes. Greater sludge bulking potential for aerobic systems. Poorer sludge dewaterability. More energy intensive.
Variations		
Aerobic	Trickling filter, rotating biological contactor, random packing upflow columns, and suspended bed upflow systems.	Oxygen pond, aerated lagoon, and activated sludge (plug flow or completely mixed regime).
Anaerobic	Random packing upflow column.	Anaerobic pond, low-rate system, high-rate upflow blanket system, fluidized bed.

Some localized areas currently regulate emissions limits for such parameters as total volatile organic substances (TVOS) and air toxics. As a rule, TVOS includes the volatile organic compounds on the priority pollutant list. Several approaches may be used to control and reduce volatile organic emissions, including source reduction, pretreatment by air or steam stripping with offgas treatment, or covering of all wastewater tanks, processes, and sewers and providing offgas treatment.

Biological Pretreatment Assessment Methodology. Two bench-scale procedures are commonly used in assessing or verifying the biotreatability of an industrial wastewater: batch treatability testing and continuous flow treatability testing. Batch testing is effective as a screening device; tests are rapid and less costly. In technology screening, a batch respirometric technique can be used to verify quickly the applicability of biological treatment to a specific waste and/or compatibility of an industrial wastewater to a municipal WWTP.

For more detailed process operation and design criteria development, continuous flow tests are appropriate. In these tests, process operation parameters (temperature, pH, mixed liquor volatile suspended solids, solids retention time, and hydraulic retention time [HRT]) are evaluated. Test periods of 2 to 4 months typically are required to develop steady-state, representative data. These data are used to perform process design and predict effluent quality in full-scale operation.

Case Histories. *CASE 1.* An on-site industrial hazardous waste (under the Resource Conservation and Recovery Act) pretreatment facility uses a batch activated sludge system to treat spent solvents, paint byproducts, and filtrate from metals precipitation before discharge to the WWTP. Parallel batch activated sludge reactors were installed because of the batch nature of discharges, 87% BOD_5 reduction being achievable through stripping and biodegradation; the variability in influent composition; space limitations; and the potential for air toxics release. Wastewater is stored and transferred in a closed system vented to an on-site vapor incinerator. Offgas from the activated sludge is vented to a chemical scrubber.

CASE 2. A specialty organic chemicals manufacturing plant was required to achieve 70% BOD reduction before discharge to a WWTP. Low-rate, anaerobic, suspended-growth pretreatment was selected based on a present-worth cost analysis of biological treatment alternatives (particularly those requiring less manpower). A pilot plant test using primary clarifier effluent from a municipal WWTP to dilute wastewater sodium concentration to < 4 000 mg/L (1:14 dilution) was performed.

CASE 3. A combination anaerobic/aerobic activated sludge pretreatment system was required for pulp and paper wastewater strictly for filamentous bulking control. Processing entailed an anaerobic zone (2.5-hour HRT) followed by a pure oxygen aerobic zone (2-hour HRT). Effluent quality of < 100 mg/L for total BOD and total suspended solids was produced.

CASE 4. Wastewater from a 246.75-g/s (0.54-lb/s) cloth bleaching and washing facility (without dyeing operations) was pretreated for discharge to a WWTP (Junkins, 1982). Wastewater from washers, bleachers, and the scour tank was collected in an equalization basin, which provided 14 hours of storage before activated sludge biological treatment. The aeration basins provided 12 hours HRT and the final clarifier operated at 2.1×10^{-4} m^3/m·s (450 gpd/sq ft). Table 9.8 summarizes the performance of this pretreatment facility.

CHEMICAL OXIDATION. Chemical oxidation processes are used to remove or transform undesirable chemical constituents to less objectionable intermediates or products. These processes involve contacting wastewater with an oxidizing agent under predetermined conditions (such as pH, catalysts, and temperature) to effect a desired reaction. The following is a typical oxidation reaction showing hydrogen peroxide oxidation of phenol to carbon dioxide and water in the presence of ferrous iron catalyst and at a reaction pH of 4:

$$C_6H_5OH + 14 H_2O_2 \rightarrow 6 CO_2 + 17 H_2O$$

Table 9.8   Bleachery wastewater pretreatment plant performance (Junkins, 1982).

Parameters[a]	Influent concentration		Effluent concentration	Discharge limit
	Design	Actual		
Flow, mgd[b]	0.12	0.09		0.12
BOD, mg/L	660	870	16	120
COD, mg/L	2 080	3 200	380	600
SS, mg/L	45	60	30	250
O&G, mg/L	34	130	20	100
pH	9	9	7	6–9
NH_3, mg/L	7	5	1	
P, mg/L	5	4	1	
Temperature, °F[c]	95	130	92	150 (Maximum)

[a] BOD = biochemical oxygen demand, COD = chemical oxygen demand, SS = suspended solids, and O&G = oil and grease.
[b] mgd × 3 785 = m^3/d.
[c] 0.555 (°F − 32) = °C.

Table 9.9 presents a list of oxidizing agents and their oxidation potentials measured in volts and in relation to the oxidation potential of chlorine: the higher the oxidation potential, the greater the reactivity and oxidation capacity of an agent. Table 9.9 also identifies some of the more commonly used oxidants in wastewater treatment (for example, ozone, hydrogen peroxide, permanganate, chlorine dioxide, and chlorine). More recent developments in oxidation processes rely on the ability of ultraviolet (UV) light irradiation to generate highly reactive hydroxyl radicles from hydrogen peroxide and ozone. In addition, irradiation can initiate molecular disruption of certain compounds (such as ring cleavage and dechlorination in aromatic compounds).

In selecting a chemical oxidation treatment process or oxidizing agent, typically desirable and sometimes competing factors are (Adams *et al.*, 1981):

- Economic feasibility, including minimal capital and operating costs;
- Reactions that will not produce undesirable secondary pollutant (for example, chlorinated organics);
- Treatment effectiveness relative to quantity of oxidizing agent used;
- Integration of the oxidation process into an existing or proposed treatment system; and
- Few site-specific constraints (such as space).

Table 9.9   Oxidation potential of various oxidizing agents.

Oxidation agent	Oxidation potential, volts[a]	Relative oxidation potential; chlorine = 1.0
Fluorine	3.03	2.23
Hydroxyl radical	2.80	2.06
Atomic oxygen (single)	2.42	1.78
Ozone[b]	2.07 (1.24)[b]	1.52
Hydrogen peroxide[b]	1.78	1.31
Perhydroxyl radical	1.70	1.25
Permanganate[b]	1.68 (0.60)[b]	1.24
Chlorine dioxide[b]	1.57 (1.15)[b]	1.15
Hypochlorous acid[b]	1.48 (0.41)[b]	1.09
Hypoiodous acid	1.45	1.07
Chlorine[b]	1.36	1.00
Bromine	1.09	0.80
Iodine	0.54	0.39

[a] Under acidic conditions; ( ) under alkaline conditions.
[b] Commonly used in wastewater treatment.

Chemical oxidation processes have been applied to the treatment of industrial wastewater to control specific pollutants of concern (for example, phenolics). However, oxidation reactions are not specific and other constituents in the wastewater will be oxidized. For example, the stoichiometric requirement for hydrogen peroxide in the oxidation of monohydric phenol is 5.06 g peroxide per gram of phenol (mole ratio of 14:1). In the treatment of coke plant wastewater, however, 12 to 16 g of peroxide per gram of phenol was required for effective treatment (Wong-Chong and Dequittner, 1980).

Additionally, organic compounds vary in their amenability to oxidation. Compounds with high reactivity are more amenable to oxidation than compounds with low reactivity. Low reactivity compounds may require greater dosages of oxidant and/or longer reaction times.

The following are examples of each type of compound (Weber, 1972).

- High reactivity—phenols, aldehydes, aromatic amines, and certain organic sulfur compounds;
- Medium reactivity—alcohols, alkyl-substituted aromatics, nitro-substituted aromatics, and unsaturated alkyl groups, carbohydrates, aliphatic ketones, acids, esters, and amines;
- Low reactivity—halogenated hydrocarbons, saturated aliphatic compounds, and benzene.

A chemical oxidation system typically consists of a contactor for mixing the wastewater and oxidizing agent; storage and feed systems, if required, for the oxidant and catalyst; pH control facilities as required; and other systems for control of flow, chemical dosages, or temperature. Chemical oxidation is greatly affected by conditions such as wastewater pH, temperature, contaminant concentration, contact time, and the presence of other oxidant-consuming constituents in the wastewater.

Chemical oxidation is widely used as a pretreatment step before biological treatment or as a posttreatment step to polish final effluents to meet regulatory criteria. Chemical oxidation, in general, is appropriate for the following circumstances:

- Concentrated waste streams with relatively low flow;
- Highly variable waste with moderate flow;
- Wastewater with constituents that inhibit or upset biological treatment processes in a WWTP;
- Wastewater that contributes to unacceptable corrosion or odors; and
- Wastewater that contains highly reactive compounds.

Hydrogen peroxide has been the most widely used oxidant in wastewater treatment, and a summary of generalized guidelines to the application of this oxidant is presented in Table 9.10. General processing characteristics for other commonly used oxidants are presented in Table 9.11.

Other oxidation processes that can be applied to the pretreatment of organic constituents in industrial wastewater include:

- Ultraviolet light irradiation combined with hydrogen peroxide and/or ozone oxidation.
- Wet air oxidation at 175 to 320°C and 2.1×10^6 to 2.1×10^7 Pa (300 to 3 000 psig). This application is limited to low-flow, high-strength hazardous wastes and sludge because of high costs.

Table 9.10   **Summary of generalized guidelines for hydrogen peroxide oxidation processing of organic compounds.**

Organic pollutant	pH	Theoretical H_2O_2 pollutant weight ratio, 100% basis	Reaction time	Catalysts
Amines	Alkaline	0.4–0.8:1	Minutes to hours	None
Aldehydes	Alkaline	0.6:1	Minutes	None
BOD/COD, TOC*	Acid	ca. 2:1	Minutes	Fe^{+2}
Hydro-quinones	Acid	4:1	Minutes	Fe^{+2}
Mercaptans, disulfides	Alkaline	5:1 mole ratio; weight depends on weight of organic component	Minutes	Chelated Fe^{+2} or Cu^{+2} generally required
Phenols, substituted phenols	Acid	94% of phenol oxidized with 2.5:1 mole ratio; 99.9% with 6:1 mole ratio; total destruction to CO_2 at 14:1 mole ratio	Minutes to hours	Fe^{+2} required; extent of phenol destruction depends on H_2O_2 phenol ratio; rate of oxidation depends on Fe^{+2} concentration

* BOD = biochemical oxygen demand, COD = chemical oxygen demand, and TOC = total organic carbon.

Table 9.11   **General process characteristics and application of chemical oxidation processes.**

Oxidant	Phase[a]	Oxidation potential, volts	Reaction condition	Influencing factor	Primary application
Hydrogen peroxide H_2O_2	L	1.78	Acidic	pH, temperature, catalyst, reaction time	Oxidation of refractory organics
Chlorine C_2	G	1.36	Acidic	pH, temperature, reaction time	Disinfection, oxidation of inorganics (CN, NH_3) Fe^{++}
HOCl	L	1.48	Acidic		
	L	0.41	Basic		
ClO_2	G	1.57	Acidic		
	G	1.15	Basic		
Ozone O_3[b]	G	2.07	Acidic	pH, temperature, catalyst, reaction time, efficient contact	Disinfection, oxidation of refractory organics and inorganics
	G	1.24	Basic		
Permanganate $KMnO_4$	S,L	1.68	Acidic	pH, temperature, reaction time	Oxidation of humic and taste and odor forming substances; oxidation of inorganics and organics
	S,L	0.60	Basic		

[a] Refers to the typical form in which the oxidants are available: G = gas, L = liquid, and S = solid.
[b] Must be generated on site.

Case Histories. *CASE 1.* A refinery wastewater was pretreated with potassium permanganate to control phenol and COD. The 0.13 m^3/s (3.0 mgd) of wastewater contained 90 to 120 µg/L of phenol and 95 to 165 mg/L of COD. Potassium permanganate processing was determined to be more cost-effective than chlorine dioxide and hydrogen peroxide.

CASE 2. A chemical manufacturing plant produced a waste stream that inhibited the performance of the plant's activated sludge biological treatment system at flow rates of 3.2×10^{-4} m³/s (5 gpm) (Linneman and Flippin, 1991). This stream contained approximately 22 000 mg/L soluble chemical oxygen demand (SCOD), 1 940 mg/L soluble biochemical oxygen demand (SBOD), and 7 500 mg/L soluble total organic carbon (STOC) and was pretreated in a ferrous sulfate-catalyzed hydrogen peroxide oxidation process. The oxidation process reduced SCOD, SBOD, and STOC by approximately 40 to 45%, and the pretreated stream was charged to the biological treatment system at flow rates of 6.3×10^{-4} to 1.1×10^{-3} m³/s (10 to 18 gpm) without inhibiting biological activity.

CASE 3. An integrated steel mill was required to further reduce phenol levels in coke plant wastewater (Wong-Chong and Dequittner, 1980). The coke plant wastewater was partially treated for ammonia and phenol by steam stripping and liquid–liquid extraction, then discharged to the plant's central treatment system, where effluent was recycled or reused in the blast furnace gas scrubber and cooling operation. Blowdown from the central treatment system was discharged to a receiving stream. A hydrogen peroxide oxidation system that used spent pickle liquor as a source of ferrous iron and acid for pH control was installed to further treat the wastewater.

PHYSICAL TREATMENT. Physical treatment technologies for processing organic constituents in industrial wastewater include

- Screening;
- Gravity separation with or without chemical coagulant (for example, ferric chloride, lime, and polyelectrolytes);
- Froth or dissolved air flotation;
- Membrane filtration;
- Adsorption; and
- Solvent extraction.

Screening, gravity separation, and froth or dissolved air flotation are discussed in Chapters 5 and 6. Membrane filtration (such as ultrafiltration, reverse osmosis, and pervaporation) are increasingly applied in supply water treatment and under special circumstances (such as when all materials of concern are in solution and flows are small). These technologies typically are cost intensive (capital and maintenance), and applications in wastewater treatment have been used primarily on secondary treatment effluents, where concerns were related to total dissolved solids removal (Nusbaum and Reidinger, 1980, and Kemmer and McCallon, 1979).

Adsorption. Adsorption processes exploit the ability of certain solid materials to preferentially concentrate specific substances from a solution onto their surfaces. This adsorption phenomenon is the result of either physical attractive forces, van der Waals forces (for example, carbon adsorption and activated alumina), or chemical adsorption where chemical bonding occurs. The latter chemical adsorption phenomenon has had little application in wastewater treatment to date. Activated carbon, a widely used adsorbent in wastewater treatment, will remove a variety of organics. Key factors affecting carbon adsorption are

- Physical characteristics of the carbon (such as surface area and pore size);
- Physical and chemical characteristics of the adsorbate (such as molecular size, molecular polarity, and chemical composition);
- Concentration of adsorbate in the liquid phase (solution);
- Characteristics of the liquid phase (for example, pH and temperature); and
- Contact time.

In determining the technical feasibility of activated carbon, adsorption tests are performed to generate adsorption isotherms. The isotherms correlate the mass adsorption capacity and equilibrium solution concentration of the adsorbate in question. From these isotherms, estimates of the theoretical carbon requirement for treatment of the wastewater in question can be made by applying the formula

$$W_{co} = [(X/M)_{co}W]/C_o \qquad (9.2)$$

Where

W_{co}	=	theoretical mass of liquid treated per unit mass of carbon,
$(X/M)_{co}$	=	adsorption capacity per unit mass of carbon at raw wastewater concentration C_o,
W	=	mass of wastewater used in the isotherm test, and
C_o	=	concentration of constituent of concern in raw wastewater.

The carbon requirement for treating a given wastewater, as determined from adsorption isotherms, typically is based on processing at static equilibrium in bench-scale batch test conditions. However, in a full-scale system, processing will occur at dynamic equilibrium based on actual contact time in the adsorption column. Consequently, more realistic process design data can be generated from adsorption tests performed in pilot plant-scale adsorption columns, where the effects of real processing parameters (such as contact times, carbon type, and particle size) can be evaluated fully. Additionally, removal of any organic is subject to interference from other materials in a mixed waste stream.

Biological treatment can be used in conjunction with powdered activated carbon (PAC) in the activated sludge tank. Difficult-to-degrade organics can be adsorbed by the carbon and/or biodegraded by bacteria on the carbon pores (Patterson, 1985). This use of carbon can also help prevent upsets from shock loads of organics that might reach the pretreatment plant. Research has investigated how combination PAC and granular activated carbon may also help increase the effectiveness of biotreatment (Adams *et al.*, 1981).

Activated carbon adsorption processing of wastewater can be implemented in a variety of process configurations. Typically, two-column, lead-lag systems are used, as illustrated in Figure 9.2.

CASE HISTORY. A chemical-manufacturing plant (Cheremisinoff and Ellerbusch, 1978) installed a dual, upflow carbon adsorption system with facilities for carbon reactivation to remove BOD, COD, phenol, and suspended solids. This system was designed to process 0.022 m^3/sec (350 gpm) at a surface loading of 1.05×10^{-3} m^3/sec/m^2 (1.55 gpm/sq ft) and a superficial retention time of 175 minutes. Each adsorber was 3.66 m diameter by 11.0 m straight side height (12 ft diameter by 36 ft), with conical bottoms and tops, and had a volume of 123 m^3 (4 343 cu ft) with a capacity for 65 830 kg (145 155 lb) of 8×3 mesh carbon. The columns operated with an actual charge of 56 296 kg (124 133 lb) carbon per absorber. Table 9.12 presents reported performance data. This treatment system included facilities for carbon regeneration, which represented significant cost to the overall treatment process and had to be factored into the overall economic feasibility.

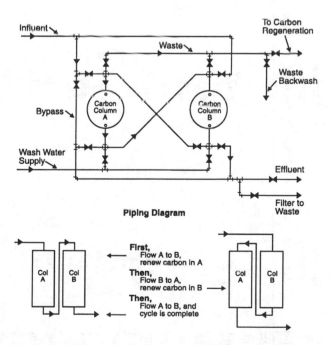

Figure 9.2 **Schematic diagram of a typical two-column carbon adsorption system.**

Table 9.12 Average performance data for the chemical manufacturing plant carbon adsorption system (Cheremisinoff and Ellerbusch, 1978).

Month	Biochemical oxygen demand	Removal, %	Chemical oxygen demand	Removal, %	Phenol	Removal, %	Suspended solids	Removal, %
September								
Influent	985		2 498		186		122	
Effluent	295	70	497	80	0.9	99.5	52	57
October								
Influent	853		2 686		197		146	
Effluent	210	75	404	85	0.5	99.7	26	82
November								
Influent	1 052		3 034		177		175	
Effluent	347	67	635	79	0.25	99.9	30	83
December								
Influent	974		2 263		193		133	
Effluent	283	71	470	79	0.3	99.8	28	79
January								
Influent	624		1 419		144		97	
Effluent	187	70	297	79	0.7	99.5	18	81
February								
Influent	806		2 207		169		102	
Effluent	227	72	343	84	0.5	99.7	17	83

Solvent Extraction. Solvent extraction is the separation of constituents of a liquid solution by contact with an immiscible liquid. The effectiveness of separation depends on the solubility distribution of the solute of concern between the two liquids. If there is some distribution difference, a degree of separation will occur, and more effective removal of the solute may require multiple contact and separation stages. The solubility distribution can be measured as an equilibrium distribution coefficient (K_D) as follows:

$$K_D = \text{Concentration of solute in solvent/concentration in wastewater}$$

The value for K_D will depend on the solvent used, temperature, and pH; the greater the values of K_D, the more effective the extraction. The K_D values for phenol extraction from water by a variety of hydrocarbon solvents are reported by Beychok (1976) and API (1969).

Figure 9.3 presents a schematic flow diagram of the solvent-extraction process. The diagram shows a single contact stage. In practice, however, this processing stage might consist of a single-stage mixing and settling unit, general mixer and multiple settling units, a single device multistage unit operating with countercurrent flows (for example, packed column or column with separation trays), or multiple sequential stages.

The processing scheme shown in Figure 9.3 includes facilities for solvent recovery and reuse. Technical applicability and cost of recovery or reuse

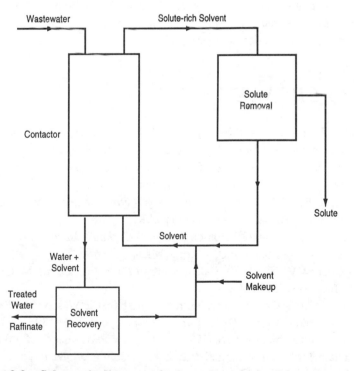

Figure 9.3   Schematic diagram of solvent extraction of wastewater.

must be considered in evaluating the feasibility of solvent extraction as a pretreatment option. Key concerns that must be considered are

- Availability and cost,
- Solubility in water,
- Potential environmental and health effects, and
- Solvent partition coefficient.

Solvent reuse typically is necessary for economic reasons but may be eliminated when solvent solubility in water is low and poses no adverse environmental or health effects.

The equilibrium partition coefficient and other pertinent process design data can be determined from batch extraction tests using various solvents and solvent-to-water ratios.

CASE HISTORY. Solvent extraction has been practiced most widely for phenol removal from coke plant, phenol manufacturing and processing, and petroleum-refining wastewaters. A coke plant was required to reduce the phenol content in its wastewater discharge to the WWTP to less than 1 mg/L (Karr and Ramanujam, 1988). The wastewater contained 400 to 1 100 mg/L phenol and was produced at an average rate of 79.5 m^3/h (350 gpm). On the basis of pilot plant testing, a 1.52-m diameter reciprocating plate column was installed. The plate stack was 18.29 m (60 ft) in height. Methylisobutyl ketone was used as the extraction solvent, and system operation produced a raffinate with phenol concentrations less than 1 mg/L.

REFERENCES

Adams, C.E., *et al.* (1981) *Development of Design and Operation Criteria for Wastewater Treatment.* Enviro Press, Inc., Nashville, Tenn.

American Petroleum Institute (1969) *Manual on Disposal of Refinery Wastes—Volume on Liquid Wastes.* Washington, D.C.

Beychok, M.R. (1976) *Aqueous Wastes From Petroleum and Petrochemical Plants.* John Wiley and Sons, London, U.K.

Cheremisinoff, P.N., and Ellerbusch, F. (1978) *Carbon Adsorption Handbook.* Ann Arbor Science, Ann Arbor, Mich.

Eckenfelder, W.W., Jr. (1970) *Industrial Water Pollution Control.* McGraw–Hill, Inc., New York, N.Y.

Junkins, R. (1982) Case History: Pretreatment of Textile Wastewater. *Proc. 37th Ind. Waste Conf., Purdue Univ.*, West Lafayette, Ind., 139.

Karr, A.E., and Ramanujam, S. (1988) Scaleup and Performance of 5 ft (1.52 m) Diameter Reciprocating Plate Extraction Column. *Solvent Extraction and Ion Exchange*, **6**, 2, 221.

Kemmer, F.N., and McCallon, J. (Eds.) (1979) Membrane Separation. *Nalco Water Handbook*, McGraw–Hill, Inc., New York, N.Y.

Linneman, R.C., and Flippin, T.H. (1991) Hydrogen Peroxide Pretreatment of Inhibitory Wastestream. Chem. Oxidation, Technol. for the Nineties, First Int. Symp., Vanderbilt Univ, Nashville, Tenn.

Nusbaum, I., and Reidinger, A.B. (1980) Water Quality Improvement by Reverse Osmosis. *Water Treatment Plant Design*. R.L. Sanks. (Ed.), Ann Arbor Science, Ann Arbor, Mich.

Patterson, J.W. (1985) *Industrial Wastewater Treatment Technology*. 2nd Ed., Butterworth–Heinemann, Boston, Mass.

Pohland, F.G. (1967) *Anaerobic Biological Treatment Processes*. Advances in Chemistry Series 105, Am. Chem. Soc., Washington, D.C.

Sawyer, C.N. (1956) *Biological Treatment of Sewage and Industrial Waters*. Vol. 1, J. McCable and W.W. Eckenfelder, Jr. (Eds.), Reinhold, New York, N.Y.

Tabak, H.H., *et al.* (1981) Biodegradability Studies with Organic Priority Pollutant Compounds. *J. Water Pollut. Control Fed.*, 53, 1503.

U.S. Environmental Protection Agency (1976) *Federal Guidelines, Pretreatment of Pollutants Introduced POTW*. Washington, D.C.

Weber, W.J. (1972) *Physiochemical Processes for Water Quality Control*. Wiley-Interscience, New York, N.Y.

Wong-Chong, G.M., and Dequittner, A.T. (1980) Physical-Chemical Treatment of Coke Plant Wastewater at Crucible Steel, Midland, PA. *Proc. Ind. Wastes Symp.*, 53rd Annu. Conf. Water Pollut. Control Fed., Las Vegas, Nev.

Chapter 10
Nutrient Removal

In biological wastewater treatment, the term *nutrient* typically refers to nitrogen and phosphorus compounds. These two constituents allow the assimilation of basic building blocks and provide mechanisms for energy transfer as organisms oxidize carbonaceous materials in wastewater treatment facilities. Also, there is some concurrent removal caused by biomass incorporation during cell growth that results from the degradation of biochemical oxygen demand.

Under specific pH and temperature conditions, nitrogen in the form of ammonia may have a toxic effect on wastewater microorganisms. Typically, however, these compounds are nonhazardous to activated sludge systems and compatible with typical municipal wastewater treatment plants (WWTP). Pretreatment is required only if the WWTP cannot effectively handle the nitrogenous load, or if the municipal discharge permit limits nitrogen and/or phosphorus, the introduction of which would cause permit violations. Experience has proven that it may be economical and beneficial to the community for nutrient removal to be done centrally. The industry and the WWTP should work together for effective solutions.

A foremost concern of wastewater treatment in the last 30 years has been to control the growth of algae in receiving waters by controlling the levels of phosphorus and nitrogen in wastewater discharges. It is important that the limiting nutrient be identified before establishing discharge requirements. Otherwise, large pretreatment expenses may not result in the desired downstream improvements.

For this discussion, the nitrogen and phosphorus forms considered to be nutrients and, therefore, subject to pretreatment scrutiny are ammonia, nitrate, and orthophosphate. Compounds like organic nitrogen, which may be toxic, are not discussed here, nor are potentially toxic forms of phosphorus such as elemental phosphorus and phosphate-incorporated pesticides.

The most frequently used pretreatment method for phosphorus removal is precipitation. For nitrogen removal, pretreatment typically consists of a nonbiological technique like ion exchange, air stripping, or breakpoint chlorination. Economic feasibility may be enhanced when nitrogen control can be combined with nitrogen recovery. Recovery of nitrogen typically is not economically viable because substantial wastewater pretreatment to remove interferences and suspended solids may be needed.

*E*FFECTS OF NUTRIENTS ON TREATMENT

PHOSPHORUS COMPOUNDS. Orthophosphate does not impair normal operation of wastewater treatment plants. Serving as a nutrient source, it passes through the plant without damaging the plant unit process. Occasional discharges of phosphate in high concentrations may form precipitates in the wastewater with calcium, aluminum, or iron compounds; these precipitates do not harm wastewater treatment plants, except to increase the maintenance effort.

NITROGEN COMPOUNDS. Ammonia. The ammonium ion (NH_4^+) in the dissociated form is relatively harmless to microorganisms. However, ammonia gas (NH_3) can be toxic to organisms in water, depending on pH and temperature. Equation 1 below shows the equilibrium reaction for the formation of ammonia gas and ammonium ion in the presence of water and hydroxide.

$$NH_4^+ + OH^- \leftrightarrows NH_3 + H_2O \tag{10.1}$$

The NH_3 gas easily diffuses through the cell wall of organisms to upset cell equilibrium and destroy the organisms. The toxicity is greater with higher pH because of increasing ammonia concentrations.

For plants using nitrification, the inhibitory effects of ammonia and nitrous acid can be detrimental to plant performance.

Most forms of organic nitrogen are hydrolyzed to ammonia through biological action. The rate of conversion from organic-nitrogen to ammonia-nitrogen influences the subsequent effects of ammonia on the bacteria in the treatment plant. With several types of industrial wastewater containing organic nitrogen (amine and coking plant wastes, for example), substantial ammonia concentration is generated in the biological system as the organic nitrogen is hydrolyzed (Adams *et al.*, 1974). Thus, a treatment facility receiving low concentrations of inorganic ammonia may generate high ammonia concentrations because of the breakdown of influent organic nitrogen. High concentrations of ammonia chelate certain heavy metals, rendering them soluble. Because this chelation may inhibit subsequent precipitation of these metals from the solution, pretreatment of the ammonia may be required.

Nitrate. Nitrate compounds may enter the municipal WWTP during the biological nitrification process, whereby ammonia is oxidized to nitrate by a specific group of microorganisms. Nitrate ions also are an excellent source of oxygen under anoxic conditions. Under these conditions, which generally prevail in the lower depths of secondary clarifiers, the nitrate ion may be reduced, releasing substantial quantities of nitrogen gas. The tiny nitrogen gas bubbles attach themselves to flocs and suspended particles and float them to the surface of the clarifier and into the effluent. This condition is especially noticeable during warm weather in nitrifying activated sludge systems with secondary clarifiers designed at overflow rates of $16 \text{ m}^3/\text{m}^2 \cdot \text{d}$ (393 gpd/sq ft) or less. The rising sludge can cause operational problems and an increase of total suspended solids in the effluent.

EFFECTS OF NUTRIENTS ON RECEIVING STREAMS

In most cases, documentation for the effect of nutrients on receiving water has been inconclusive. In many natural water systems, growth limitations for algae result from increased nutrient levels. Allowable nitrogen (nitrate [NO_3^-], nitrite [NO_2^-], ammonia [NH_3], and total Kjeldahl nitrogen [TKN]) and phosphorus concentrations have been as low as 5 and 1 mg/L, respectively. Other standards for nitrogen refer to more classical conditions of oxygen sag, wherein ultimate oxygen demand is defined as a factor times TKN, which reflects the oxygen requirements for complete nitrification. Studies to determine the limiting factor for nuisance algal growth can be expensive and time consuming. Nutrient control without this understanding, however, may not achieve the desired result.

PRETREATMENT PROCESSES

Pretreatment processes may be biological and physical–chemical. Because of the limited efficiency of nitrogen and phosphorus removal in biological facilities, biological pretreatment is generally not used for nutrient removal before discharging to a municipal biological WWTP. Biological treatment of nutrients at WWTPs is being done, however. Pretreatment for phosphorus removal is considered technically feasible using physical–chemical methods.

PHOSPHORUS REMOVAL. Coagulation-precipitation typically offers the most practical method of phosphorus removal. However, this method requires chemical addition and also results in substantial sludge production. For precipitation of phosphorus, three cations—iron, aluminum, and calcium, in the forms of ferric chloride, alum, and lime, respectively—have been used most often (Table 10.1). Ferrous chloride and ferric sulfate also have been used, however. The equipment required for the precipitation is basically the same for any of these chemicals, that is, chemical addition, rapid mix, flocculation, and clarification (Figure 10.1). The need for a flocculator varies with the design and operation of the clarifier. The selection of a particular chemical depends on availability and cost, initial pH of the raw wastewater, resulting sludge production, and projected method of sludge disposal.

The optimum pH for removal of phosphorus is shown in Figure 10.2. This figure also shows some of the phosphate precipitates that form. The curves show a simplified chemical equilibrium system not truly representative of the actual precipitates. DeBoice and Thomas (1975) and others have conclusively shown that apatite $(Ca(PO_4)OH)$ is frequently oversaturated and that its solubility is generally not the controlling reaction for phosphorus precipitation. DeBoice reported that phosphorus solubility is most accurately predicted by the empirical equation

$$(Ca)^3 (PO_4)^2 = 10^{-25.46} \tag{10.2}$$

Where Ca is the concentration of the calcium ion and PO_4 is the concentration of the phosphate ion. This equation indicates that the controlling precipitate is tricalcium phosphate $(Ca_3(PO_4)_2)$ and that the solubility product constant pK is 25.46.

Another factor that influences phosphorus precipitation is the coprecipitation of calcite $(CaCO_3)$, which substantially increases the amount of lime and the resulting quantity of sludge produced. The effect of calcite coprecipitation is indicated by the flat shape of the curve on Figure 10.2 in the pH range of 8.8 to 10.0. In this range, the phosphorus concentration actually increases slightly because of competition for the calcium ion by calcite precipitation.

Table 10.1   Chemical coagulants for phosphorus precipitation.

Chemical	Dosage range, mg/L	pH	Comments	Settling, $m^3/m^2 \cdot d$	Chemical sludge production, mg/L
Lime	150–500	9.0–11.0	For colloid coagulation and phosphorus removal in wastewater with low alkalinity and high and variable phosphorus	45	500
Alum	75–250	4.5–7.0	For colloid coagulation and phosphorus removal in wastewater with high alkalinity and low and stable phosphorus	30	50
$FeCl_3$	35–150	4.0–7.0	For colloid coagulation and phosphorus removal in wastewater with high alkalinity and low and stable phosphorus	30	45
$FeSO_4 \cdot 7H_2O$	70–200	4.0–7.0	Where leaching of iron in the effluent is allowable or can be controlled and where economical source of waste iron is available (for example, steel mills)	30	45

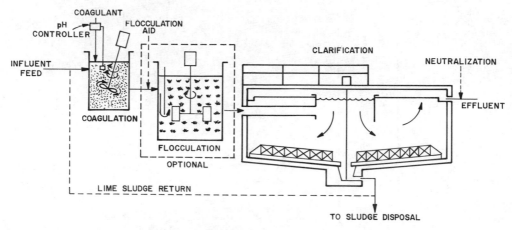

Figure 10.1   Three-stage chemical treatment system.

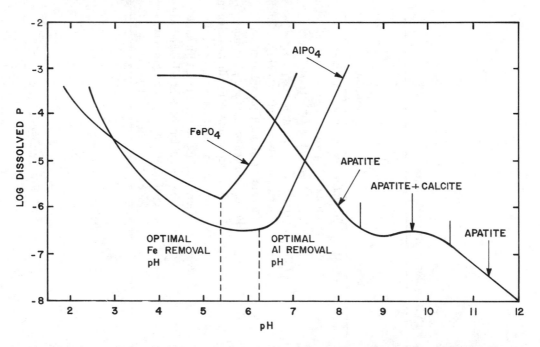

Figure 10.2   Equilibrium solubility diagram for ferric, aluminum, and calcium phosphate precipitates. Note that Al and Fe diagrams are for solutions in equilibrium with indicated precipitate; Ca diagram is for precipitation from system; Ca = 10^{-3} M; P = 5×10^{-3} M (Jenkins *et al.*, 1970).

Typically, lime is the most economical chemical to use. However, if the wastewater pH is low (less than 6.5 to 7.0), substantial quantities are required to achieve the alkalinity needed for lime precipitation, and large volumes of sludge result. Ferric chloride and alum coagulate phosphorus most effectively at pH 5 to 8 with lower sludge production than lime.

If landfilling is the sludge-disposal method selected, any of the three co-agulants are satisfactory because all produce a stable, disposable sludge. However, if land-spreading, land-spraying, or land-farming applications are used, lime yields the most readily acceptable biosolids. Alum and ferric biosolids tend to clog the soil and thus limit the application rate and longevity of the disposal area.

Another promising technology is the biological removal of phosphorus. This is accomplished by "stressing" the biological organisms by placing them in an anoxic zone. When biological conditions are optimal, "luxury uptake" and concentration of the phosphorus into a cell where it can be wasted is accomplished. Two notable examples of this technology are the Fayetteville, Arkansas, and Hampton Roads, Virginia, WWTPs.

NITROGEN REMOVAL. Treatment for removal of nitrogenous compounds is complex and costly. Treatment can result in byproducts that may cause air pollution problems because of ammonia released to the atmosphere or highly alkaline solution condensation deposits on the immediately surrounding area. Elaborate subsequent disposal techniques may be required for the salt or chemical brines. Consequently, the pretreatment process selected for removing or reducing nitrogenous compounds should be investigated thoroughly and consider existing technology and possible reuse of the byproducts generated.

Nitrogen occurs in three forms that may require treatment: ammonia, nitrate, and organic nitrogen. Only ammonia and nitrate are considered in pretreatment because the organic nitrogen may be converted to ammonia in biological systems. The basic processes available for removal of ammonia or nitrate include

- Ammonia—biological nitrification (ammonia biologically oxidized to nitrate); air or steam stripping (ammonia stripped into atmosphere or recovered in a spent alkaline solution); ion exchange (ammonia selectively removed from wastewater and subsequently disposed of as spent regenerant); and breakpoint chlorination (ammonia converted to nitrogen gas).
- Nitrate—biological denitrification (nitrate biologically converted to nitrogen gas) and ion exchange (nitrate selectively removed from wastewater and disposed of as a spent regenerant).

In most cases, pretreatment for nitrogen removal involves a physical-chemical process, that is, air stripping, ion exchange, or breakpoint chlorination. Biological pretreatment typically would be redundant with the municipal WWTP. Rare cases, such as in a munitions or fertilizer plant, may dictate biological pretreatment because of the excessively high levels of ammonia. Because of WWTP sewer fees, however, the industry may wish to examine either direct discharge to a receiving water or recycling of the treated wastewater.

Air or Steam Stripping of Ammonia. Air or steam stripping processes for ammonia removal involve raising the wastewater pH to high levels (pH 10.5 to 11.5) and providing sufficient gas–water contact to strip the ammonia gas from solution (Adams, 1971). Steam is efficient for stripping because of the low partial pressure (concentration) of ammonia in the stripping gas, elevated temperatures, and absence of other gases in the steam. Because of energy considerations, however, steam stripping is more likely to be feasible when waste steam is available from another process at the facility.

The four important considerations for stripping ammonia are elevated pH levels, minimal surface tension of the air–water (or steam–water) interface, large volumes of air or steam, and ambient operating temperature.

As the pH of the wastewater is increased above 7, the equilibrium (see Equation 10.1) is shifted toward ammonia gas that can be removed from the liquid by causing sufficient air–water contact.

After the ammonium ion has been converted to ammonia gas, two major factors affect the rate of transfer of this gas from the wastewater to the surrounding atmosphere: the surface tension of the air–water interface and the driving force resulting from the difference in ammonia concentration in the water and the surrounding air as measured by Henry's law. Repeated droplet formation, rupture, and reformation greatly enhance ammonia stripping operations by reducing surface tension and increasing ammonia outgassing. To maintain a sufficient driving force between the liquid and gas phase, it is necessary to reduce the concentration (partial pressure) of ammonia in the gas phase by circulating large quantities of air or steam rapidly through the water droplets. The same principles of droplet formation and reformation and the necessity of large gas–liquid requirements apply to conventional cooling towers and explain the adaptability of these towers to the ammonia stripping process.

DESIGN BASIS. For design of an air stripping tower, a gas-to-liquid ratio and hydraulic loading rate of liquid to the top of the stripping tower must be selected. Common ranges for these design parameters are (Adams, 1971, and Culp and Culp, 1971):

- Required gas-to-liquid ratio—2 to 4 kg air/kg water (2 to 4 lb/lb).
- Treated hydraulic loading—0.3 to 2 L/m²·s (0.4 to 2.8 gpm/sq ft).

Other design parameters are

- Air pressure drop—0.1 to 0.3 kPa (0.4 to 1.2 in.) water.
- Fan-tip speed—46 to 61 m/s (9 057 to 12 011 ft/min).

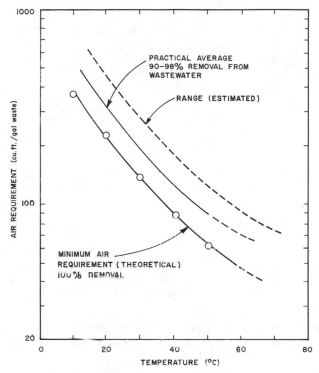

Figure 10.3 **Temperature effects on gas–liquid requirements for ammonia stripping (cu ft/gal × 7.482 = m³/m³) (Tchobanoglous, 1970).**

- Fan motor—1- or 2-speed.
- Packing depth—6 to 8 m (20 to 26 ft).
- Vertical and horizontal packing spacing—50 to 100 mm (2 to 4 in.).

The gas-to-liquid requirements are a function of temperature and are approximated as shown in Figure 10.3. The design of the media can improve the formation of water droplet surface area and, thus, stripping efficiency. The gas head loss is also a function of media design. At hydraulic loadings greater than $2/L/m^2$·s (2.8 gpm/sq ft), a sheeting effect occurs in the stripping tower and droplet formations are retarded, thereby inhibiting the stripping process. This phenomenon is illustrated in Figure 10.4.

OPERATIONAL CONSIDERATIONS. Apart from economic considerations, the major advantage of ammonia stripping is the ability to control the process for selected ammonia removals. The operator can adjust the pH upward or downward to achieve greater or lower ammonia removals. However, several

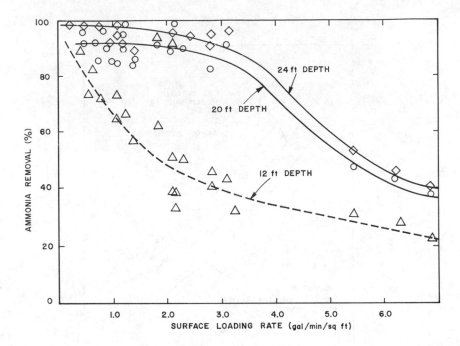

Figure 10.4 **Effects of hydraulic loading on ammonia removal at various depths (gpm/sq ft × 0.679 02 = L/m^2· s; ft × 0.304 8 = m) (Tchobanoglous, 1970).**

disadvantages are inherent in the air stripping process through conventional cooling towers. They include the following (Adams, 1971):

- Cold weather operation—increased ammonia solubility with lower temperature, tower's inability to operate at wet bulb temperatures less than 0°C (32°F), and occurrence of fogging and icing problems;
- Deposition of calcium compounds within the tower, resulting from pH adjustment using lime and the subsequent reaction of the solution with the carbon dioxide in the stripping air;
- Odors and air pollution controls requirements, especially in sensitive air quality basins;
- Reduction in pH and subsequent removal efficiency because of carbon dioxide absorption;
- Deterioration of wood material caused by delignification as a result of the high pH of the wastewater;
- Deposition of high pH solutions caused by condensation after stripping; and
- Altitude effects caused by a reduction of partial pressure of the stripping air.

In northern climates with extreme winter temperatures, problems may result from freezing of the tower when the wet bulb temperature is less than 0° (32°F). Also, ammonia is difficult to transfer from solution because of its increased solubility at low temperatures. In larger facilities, vaporization of a warm influent waste causes extreme fogging conditions with the possibility of ice forming on nearby roads or buildings.

Many municipal WWTPs use lime to increase the pH and alkalinity of the wastewater. Beside increasing the pH for ammonia removal, lime also aids in precipitation of phosphorus compounds, a desirable objective in most municipal WWTPs. The supernatant from the pH adjustment and precipitation process typically is supersaturated with calcium compounds that can precipitate on surfaces such as the tower medium.

Deposits of calcareous materials are a major problem with high pH ammonia stripping. Periodically, the tower must be shut down and the media cleaned, either manually or with high-pressure water jets. The presence of these calcium deposits also has been noticed in piping systems downstream from the tower. This condition results from a phenomenon similar to the one that causes these materials to precipitate in the stripping tower. Plastic media may facilitate cleaning operations because of their ability to withstand acid washes.

A significant consideration in the application of conventional cooling towers for ammonia stripping is the type of media. In the past, redwood slats were used as a standard packing media. Because the high pH wastewater required for ammonia stripping has resulted in delignification and subsequent deterioration of the wood packing, plastic media are currently used in many stripping towers.

In the case of industrial wastes, where the ammonia concentration of the wastewater may be extremely high, an air emission problem may develop causing odors or requiring an air permit for the discharge of large quantities of ammonia to the atmosphere. In these instances, if the ammonia concentration in the air is significant, the air stream from the tower may be scrubbed through a dilute sulfuric acid solution to recover the ammonia (Figure 10.5). A system for reclaiming the ammonia with sulfuric acid also helps eliminate other operational problems associated with conventional stripping. Nevertheless, even with these limitations, ammonia stripping has been shown to be a feasible, reliable method for ammonia removal under the right climatic conditions and with the proper precautions taken for scale prevention or removal.

The first operating ammonia-recovery system, although not an industrial pretreatment system, was installed at Tahoe-Truckee Sanitary District (Truckee, California). To date, the system, which uses clinoptilolite for ammonia removal and sulfuric acid for ammonia recovery, has worked satisfactorily and provided an ammonium sulfate solution of approximately 10%. This recovered ammonium sulfate is given away to a local fertilizer manufacturer.

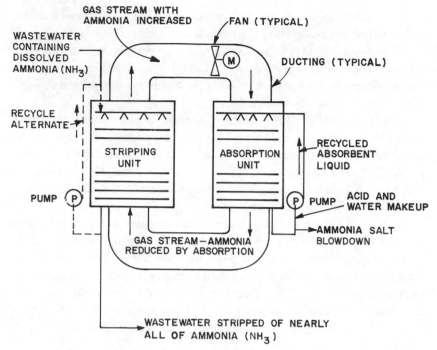

Figure 10.5 **Process for ammonia removal and recovery (U.S. EPA, 1974).**

Other Stripping Options. Studies indicate limited stripping of ammonia in high-pH, surface-agitated ponds (U.S. EPA, 1974); however, the process is energy intensive because of inherent deficiencies of high-speed aerators. Where adequate land exists and only 30 to 50% ammonia removal is required, these ponds may constitute a reliable process.

Other possible methods of ammonia stripping from solution include use of diffused-air and mechanical surface aerators. Recently, a helix-type aerator has been used with varying degrees of success. The major advantage of compressed air aerators is their ability to maintain the wastewater at a higher temperature during stripping. However, these devices are not capable of supplying the extremely high air quantities as economically or efficiently as countercurrent-flow stripping towers.

Ion Exchange for Ammonia and Nitrate Removal. Ion exchange is a process in which ions, held by electrostatic forces on the surface of a solid, are exchanged for environmentally benign ions. Although natural zeolites such as alumino-silicates have been used in the past, their low capacities and high attrition rates typically have rendered them inefficient in wastewater treatment. Most ion exchange resins in use today are synthetic materials consisting of a network of hydrocarbon radicals with soluble ionic functional groups attached. The nature of the groups determines the resin behavior, and the number of groups per unit weight of resin determines the exchange capacity. The group type affects the selectivity and exchange equilibrium of the resin.

Clinoptilolite, a naturally occurring zeolite, is highly selective for the ammonium ion and has been used frequently for ammonia removal. Regeneration can be accomplished with salt or caustic soda. Caustic soda offers the advantage that the ammonia can be easily stripped and the solution can be reused. However, the experience at Tahoe-Truckee Sanitary District indicates that calcareous compound depositions still are a significant problem in the stripping tower, and that regenerant must be softened with soda ash to precipitate calcite before being introduced to the stripping tower. Moreover, clinoptilolite is subject to attrition and has limited adsorptive capacity. Finally, ion exchange as a pretreatment process is more economically feasible than other processes only if the spent regenerant can be reused or sold.

SYNTHETIC RESINS SYSTEM. A few full-scale pretreatment plants in operation were designed specifically to remove and recover both ammonia and nitrate from industrial wastewater using ion exchange. The following example discusses the design of a system for pretreatment and recovery of ammonium nitrate (see Figure 10.6).

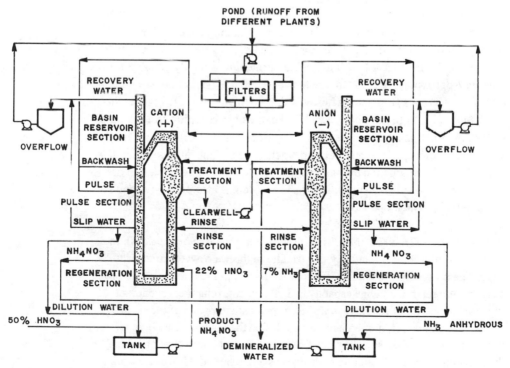

Figure 10.6 Ammonium nitrate recovery system.

The system was designed to remove ammonium nitrate (429 mg/L ammonia-nitrogen and 214 mg/L nitrate-nitrogen) and produce a regenerant suitable for recovery at approximately 15 to 20% ammonium nitrate (Bingham, 1972). The following steps summarize the process:

1. Collection of all plant wastewater, including runoff, in large settling ponds.
2. Filtration through dual media filters to remove particulate matter.
3. Removal of ammonia-nitrogen from a concentration of 429 mg/L to 3 mg/L using a weak base (hydroxide resin).
4. Regeneration of the cation resin with 22% nitric acid. (The concentration of this solution was found to be the optimum both to prevent repeated attrition for continued resin swelling and shrinkage and elute 99% of the ammonia from the resin during regeneration.)
5. Regeneration of the anion resin with 7% ammonium hydroxide. (The hydroxide ions displace the nitrate, thereby providing the concentrated ammonium nitrate regenerant.)
6. Recovery of ammonium nitrate in 18% solution by combining the spent regenerants from each separate unit and neutralizing the excess acid with ammonium hydroxide.
7. Production of a 30% marketable ammonium nitrate solution by combining the 18% ammonium nitrate spent regenerant with an 83% ammonium nitrate solution from the production facility.

The results of full-scale operation of this system indicate excellent removal of ammonium nitrate and the production of a regenerant of 20% total solids containing at least 18% ammonium nitrate that is highly suitable for recovery. Approximately 5 400 metric tons/a (6 000 tons/yr) nitrate are produced; 50% comes from the wastewater, the other half from the regenerants.

Historically, there have been accidental explosions resulting from the use of the combined ion-exchange system. These accidents typically have resulted from improper operation during regeneration. Concentrated nitric acid exposed to the organic ion exchange resin results in the rapid oxidation of this material and subsequent explosion. The system is also expensive and requires a well-trained operator.

CLINOPTILOLITE SYSTEM. Clinoptilolite, found in large quantities in the northwestern U.S., is a relatively new zeolite in the wastewater treatment field. The most important considerations for its use in the ion-exchange process include resin-exchange capacity, pH effects on exchange, regenerant type, optimization of regenerant, and attrition characteristics of the resin.

After exchange of ammonia on clinoptilolite, the resin may be regenerated with either calcium or sodium salts. One advantage of using clinoptilolite is the inexpensiveness of calcium hydroxide as a regenerant. However, work by Koon and Koffman (1971) indicated that sodium salts are more effective than calcium salts in regeneration and retaining ammonia capacity in the resin. Experience has shown that a mixture of sodium and calcium salts provides the most economical and effective regenerant for ammonia elution from the resin.

Optimum conditions for exhaustion of the resin occur at pH 4 to 8. Operation outside of this range results in a rapid decrease of ammonia exchange capacity and increased ammonia leakage before the beginning of breakthrough. Typically, the optimum pH for exchanges ranges from 6 to 7.

Optimizing regenerant use involves an evaluation of regenerant flow rate, strength, pH, and quantity. *Efficiency of regeneration*, a term that applies to optimum regenerant use, is defined in the following equation (Koon and Koffman, 1971):

$$\text{Efficiency (\%)} = (\text{NH}_3 \text{ removed during previous exhaustion, equivalent/Na used for regeneration, equivalents}) \times 100 \qquad (10.3)$$

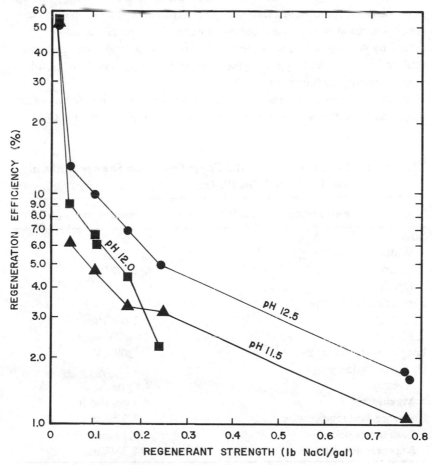

Figure 10.7 **Clinoptilolite regenerant concentration effects on regeneration efficiency (lb/gal × 119.8 = kg/m³).**

The results of clinoptilolite regeneration expressed as percent efficiency are shown in Figure 10.7. The efficiencies for pH 12.5 were about twice those for pH 11.5. The efficiencies observed are considerably lower than those observed in ion-exchange, water-softening operations. However, the efficiencies shown in Figure 10.7 are based on an ammonia elution of 95% and do not reflect the elution of other ions that would increase the overall regeneration efficiency.

The attrition rates of clinoptilolite increase with increasing pH. Experience indicates that the attrition rates are approximately 0.25, 0.35, and 0.55% per cycle using regenerants at pHs of 11.5, 12.0, and 12.5 respectively.

A basis for design is given by the criteria used for the Upper Occoquan Sewage Authority in Virginia (Table 10.2). The two most difficult operation problems in an ion-exchange system arise from fouling of the resins and excessive attrition rates. This conclusion assumes that adequate pretreatment exists to prevent clogging of the resins with suspended solids. Fouling of resins may occur with precipitated salts and with excessive organic loadings to the resins that result in subsequent organic fouling. These problems can be solved by pretreatment ahead of the resin to remove the materials causing fouling, or by periodically rinsing the resin with a strong acid, base, or solvent to remove the fouling materials.

Although a highly concentrated regenerant results in greater elution of the ammonia from the resin, high concentrations cause a resin to shrink. On

Table 10.2 Design criteria for the Upper Occoquan Sewage Authority, Virginia (U.S. EPA, 1974).

Parameter	Design value
Size of bed	
Width	10 ft[a]
Length	40 ft
Depth	4 ft
Service cycle loading	
Average	9.1 BV[b]/hour
Maximum	14.1 BV/hour
Length of service cycle	200 BV
Hydraulic loading	
Average	4.4 gpm/sq ft[c]
Maximum	6.9 gpm/sq ft
Exchange-bed regeneration	3.1 hours
Length of cycle	3.1 hours
Regeneration rate	10 BV/hour

[a] ft × 0.304 8 = m.
[b] BV = bed volume.
[c] gpm/sq ft × 0.679 02 = L/m^2·s.

subsequent exposure of the resin to the more dilute waste stream after regeneration, the resin tends to swell. This constant shrinking and swelling causes the resin to crack and substantially increases the rate of attrition. Therefore, the operator must increase regenerant strength after each cycle to achieve maximum elution of the constituents from the resin and reduce the shrinking–swelling phenomenon.

Other operation problems also develop when using a high-pH regenerant. Plugging of the zeolite bed occurs with the formation of magnesium hydroxide and calcium carbonate. Excessive backwashing, which is required to remove these deposits, accelerates the attrition rates (U.S. EPA, 1974).

Breakpoint Chlorination. Breakpoint chlorination is the addition of chlorine to dilute aqueous solutions to chemically oxidize ammonium ions to products—primarily nitrogen gas. Many complex reactions are involved, and the success of the process depends on proper application of chemical and design techniques. Under proper operating conditions, 95 to 99% of the ammonia nitrogen in wastewater can be converted to nitrogen gas (U.S. EPA, 1975). This process can be optimized operationally so that no breakpoint reaction intermediate compounds like monochloramine or dichloramine are detected. Typically, the ammonia-nitrogen fraction that is not converted to nitrogen gas consists of nitrate and nitrogen trichloride.

The oxidation of ammonia by chlorine to the final nitrogen gas product is a complex process with many competing reactions, as noted below. Nitrate and chloramines (which are referred to as "combined residual" as opposed to "free residual") are formed during the process.

$$Cl_2 + (HCO_3)^- \leftrightarrows HOCl + CO_2 + Cl^- \qquad (10.4)$$

$$(NH_4)^+ + HOCl \leftrightarrows NH_2Cl \text{ (monochloramine)} + H_2O + H^+ \qquad (10.5)$$

$$NH_2Cl + HOCl \leftrightarrows NHCl_2 \text{ (dichloramine)} + H_2O \qquad (10.6)$$

$$(NH_4)^+ + 4(HOCl) \leftrightarrows (NO_3)^- + 6H^+ + 4Cl^- + H_2O \qquad (10.7)$$

$$NHCl_2 + HOCl \leftrightarrows NCl_3 \text{ (nitrogen trichloride)} + H_2O \qquad (10.8)$$

$$2(NH_4)^+ + 3Cl_2 + 8(HCO_3)^- \leftrightarrows N_2 + 6Cl^- + 8CO_2 + 8H_2O \qquad (10.9)$$

The type of reaction and the extent of its predominance depend on certain process variables such as pH, temperature, contact time, and the initial chlorine-to-ammonia-nitrogen ratio.

The theoretical breakpoint curve is illustrated in Figure 10.8. In Zone 1 of Figure 10.8, the major reactions are between chlorine and ammonia (Equation 10.5). The peak of the breakpoint curve theoretically occurs at a 5:1 weight ratio of chlorine to ammonia-nitrogen (a molar ratio of 1:1). In Zone 2, conditions favor the formation of dichloramine (Equation 10.6) and the oxidation of ammonia as shown in Equation 10.7. At the breakpoint, where the theoretical ratio of chlorine to ammonia-nitrogen is 7.6:1.0, the ammonia concentration is at a minimum. After the breakpoint, the accumulation of free chlorine residual as well as small quantities of dichloramine, nitrogen trichloride, and nitrate result (Equations 10.6 and 10.8). The optimum pH for breakpoint chlorination seems to be within the range of pH 6 to 7. Attempts at breakpoint chlorination outside this range have indicated an appreciably higher chlorine requirement and slower reaction rate for breakpoint.

Two conclusions can be made concerning pH sensitivity with respect to the formation of nitrate and nitrogen trichloride. First, the pH adjustment chemical should be added to the chlorine solution before application of the chlorine solution to the process. Effective mixing of the chemicals is important, because a disproportionate mixture of chlorine solution and pH adjustment chemical can occur and may result in pockets of liquid in the breakpoint reaction zone that are not at the desired pH. If the breakpoint reaction occurs in these pockets, excessive concentrations of nitrate as nitrogen

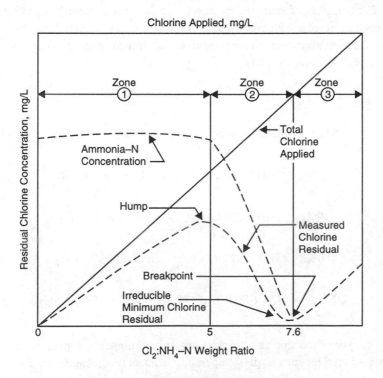

Figure 10.8   Theoretical breakpoint chlorination curve.

trichloride result. Secondly, the operating pH for breakpoint chlorination should be maintained at 7. This pH results in optimum chlorine-to-ammonia ratios and the fastest reaction rates, thus reducing nitrogen trichloride (NCl_3) production to 0.2 mg/L or less and reducing attendant odor generation.

In a system designed for breakpoint chlorination of industrial wastewater, it is essential that pretreatment reduce high concentrations of inorganic or organic compounds (sulfides, sulfites, thiosulfites, ferrous ions, phenols, amino acids, proteins, and carbohydrates). These compounds can exert excessive chlorine demand on the process.

Sufficient hydraulic or mechanical energy must be provided to ensure rapid and thorough blending of the chlorine solution, pH adjustment chemical, and wastewater. Adequate mixing ensures process consistency, a prerequisite for the feedback type of instrumentation used in the process control. After the chemicals are mixed, the reaction rate proceeds rapidly, and a 1-minute contact period is sufficient for full-scale applications. The contact basis should approach plug flow conditions as closely as possible.

Automatic process control is essential to reduce formation of NCl_3 and NO_3^-. Because of NCl_3 provision, sufficient ventilation must be provided if the reaction basin is enclosed.

Through the hydrolysis of chlorine gas in solution and the oxidation of ammonia, acidity is generated. Stoichiometrically, 14.3 mg/L of alkalinity, as calcium carbonate, is consumed for each milligram per litre of ammonia nitrogen oxidized to nitrate. In practice, approximately 15 mg/L of alkalinity is consumed per milligram per litre of ammonia oxidized. Consequently, if the ammonia concentration is significant, the alkalinity must be substantial to provide buffering capacity and maintain pH at a reasonable level.

In performing breakpoint chlorination, free residual of chlorine that is generated may require removal with a reducing compound before discharge to a municipal WWTP.

Because of the significant weight ratio of chlorine to ammonia required (minimum of 7.6:1.0), the resulting operating cost is substantially higher than costs for the other alternatives. In general, breakpoint chlorination is justifiable only if capital cost must be reduced, high operating costs can be tolerated, and ammonia concentrations are low. A breakpoint system can be installed more rapidly and for a lower capital cost than any other system, and sometimes can be used as a "quick fix."

Finally, during breakpoint chlorination, chlorinated organic compounds generated from some applications may be toxic or pass through the municipal WWTP and result in the discharge of toxicity to receiving water. In those cases where ammonia concentrations are greater than 10 to 15 mg/L and certain organic compounds are present, the breakpoint chlorination process typically should not be selected.

REFERENCES

Adams, C.E. (1971) Theoretical and Practical Consideration in the Design of Ammonia Stripping Towers. Paper presented at Environ. Assoc. Program, Vanderbilt Univ., Nashville, Tenn.

Adams, E.E., *et al.* (1974) Treatment of Two Coke Plant Wastewaters to Meet EPA Effluent Criteria. *Proc. 29th Annu. Ind. Waste Conf., Purdue Univ.*, West Lafayette, Ind.

Bingham, E.C. (1972) Closing the Loop on Wastewater. *Environ. Sci. Technol.*, **6**, 8, 692.

Culp, R.L., and Culp, G.L. (1971) *Advanced Wastewater Treatment*. Van Nostrand Reinhold, New York, N.Y.

DeBoice, J., and Thomas, J. (1975) Chemical Treatment for Phosphate Control. *J. Water Pollut. Control Fed.*, **47**, 9, 2246.

Jenkins, D., *et al.* (1970) Chemical Processes for Phosphate Removal. Wastewater Acclimation and Reuse Workshop, Coll. of Eng. and School of Public Health, Univ. Calif., Lake Tahoe.

Koon, J.H., and Koffman, W.J. (1971) *Optimization of Ammonia Removal by Ion-Exchange Using Clinoptilolite*. Water Pollut. Control Reservoir Serv. No. 17080, DAR 09, Washington, D.C.

Tchobanoglous, G. (1970) Physical and Chemical Processes for Nitrogen Removal: Theory and Application. Presented at 12th Sanit. Eng. Conf., Univ. Ill.

U.S. Environmental Protection Agency (1974) *Physical, Chemical, Nitrogen Removal: Wastewater Treatment*. Technol. Transfer Seminar Publ., Washington, D.C.

U.S. Environmental Protection Agency (1975) *Process Design Manual for Nitrogen Control*. Technol. Transfer, Seminar Publ., Washington, D.C.

Index

A

Acidic agents, 139
  carbon dioxide, 139
  sulfuric acid, 139
Acidity, definition, 127
Activated sludge, 38
  inhibition levels, 9
  organics removal, 207
Adsorption, 35, 180
  organics, 217–220
Aerated lagoons, 38
Aeration, 35, 61
Aerobic, interferences, 105
Aerobic treatment, organics, 209
Air/steam stripping, ammonia
    removal, 232–235
Alkalinity, definition, 127, 128
Alternating flow diversion, 47, 50
Ammonia/nitrate removal, ion
    exchange, 236
Ammonia removal
  air/steam stripping, 232–235
  breakpoint chlorination, 241–243
  other stripping options, 235
Anaerobic digestion
  inhibition levels, 12
  recovered oil, 121
Anaerobic interferences, 105
Anaerobic treatment, organics, 209
Analysis
  cost, 41
  fats, oil, and grease
    floatable, 110
    total, 109
    total petroleum hydrocarbons,
      109

B

Baffling, 60
Basic agents
  caustic soda, 138
  lime, 135
  magnesium hydroxide, 139
  sodium bicarbonate, 138
  sodium carbonate, 138
Biological pretreatment, 37
Biological treatment
  activated sludge, 38
  aerated lagoons, 38
  organics, 194
    assessment methodology, 211
    case histories, 211
    process configuration, 195
      priority pollutants, 196–207
  packed-bed reactors, 38
  rotating biological contactors, 38
  trickling filters, 38
Biosolids, 99
Breakpoint chlorination, ammonia
    removal, 241–243
Buffering capacity, definition, 129

C

Categorical pretreatment standards,
  18
Categorical industrial users, 21
Centrifugation, 87
  fats, oil, and grease, 118
Chemical coagulation, 74

I

Incineration, 98
 recovered oil, 120
Industrial wastewater
  characteristics, 3, 190
Industrial user, 17
 categorical, 21
Industrial wastewater processes, 34
Industries
 heavy metals removal, 156
 metals occurrence, 157, 158
Industry categorical pretreatment
  standards, 18
Inhibition levels
 activated sludge, 9
 anaerobic digestion, 12
 nitrification, 11
 trickling filter, 10
Intermittent flow diversion, 50
Ion exchange, 37, 177–180
 ammonia and nitrate, 236
  clinoptilolite, 238–241
  synthetic resins, 237

L

Land application, biosolids, 99
Landfarming, 97
Landfill, recovered oil, 120
Landfilling, 97
Local limits, 16

M

Management, wastewater, 29
Management strategies, 27
 pollution prevention, 27
 pretreatment, 33
 waste minimization, 27
Manufacturing, wastewater
  management, 29
Mass balances, 30

Mechanical mixing, 60
Metalworking, fats, oil, and grease,
  110
Mixed combined flow, 47, 54
Mixed fixed flow, 48
Mixing
 mechanical, 60
 requirements, 60

N

Neutralization, 36, 123
 alternative agents, 133
  acidic agents
   carbon dioxide, 139
   sulfuric acid, 139
  basic agents, 135
   caustic soda, 138
   lime, 135
   magnesium hydroxide, 139
   sodium carbonate, 138
   sodium bicarbonate, 138
  chemical, properties, 137
  factors, 136
 batch-control systems, 143
 bulk storage requirements, 140
 continuous flow systems, 145
 cost, acidity and alkalinity, 140
 definitions
  acidity, 127
  alkalinity, 127, 128
  buffering capacity, 129
  pH/pOH, 126
 design, pH-control systems, 142
 operational considerations, 149
  process control, 151
  raw materials, 149
 wastewater characterization, 129
  flow and load variability, 132
  sludge production, 133
  titration curves, 129–131
Nitrate/ammonia removal, ion
  exchange, 236
Nitrification inhibition levels, 11
Nitrogen compounds, 226

Noncategorical significant
industrial user, 21
Notifications, hazardous waste, 22
Nutrient removal, 225–243
ammonia, 232–235
air/steam stripping, 232–235
breakpoint chlorination,
241–243
other stripping options, 235
ammonia/nitrate, ion exchange,
236
clinoptilolite, 238–241
synthetic resins, 237
nitrogen, 231
pretreatment processes, phosphorus removal, 228–231
streams, effects, 227
treatment effects, phosphorus
compounds, 226

O

Oil, recovered, 119
Oil removal, 103
Oxidation-reduction, 37

P

Packed-bed reactors, 38
Passthrough, 106
Pesticides, 6
Petroleum, fats, oil, and grease, 112
pH
definition, 126
design control systems, 142
Phosphorus, nutrient removal,
228–231
Phosphorus compounds, 226
Physical separation, 34
adsorption, 35, 217–220
aeration, 35
dissolved air flotation, 35
electrodialysis, 36
filtration, 35

flow equalization, 35
organics, 217
adsorption, 217–220
solvent extraction, 221, 222
pretreatment, 34
reverse osmosis, 36
screening, 35
sedimentation, 35
ultrafiltration, 36
Plate settler, 175
pOH definition, 126
Pollutants
cross-media, 39
pesticides, 5
active ingredients, 6
pretreatment standards, 4
priority, 4
metals, 4
semivolatile organics, 5
volatile organics, 4
Precipitation, 37
Presses
belt, 88, 89
recessed-plate, 88
screw, 89, 90
Pretreatment, 33
biological, 37
organics
aerobic, 209
anaerobic, 209
chemical, 36
cost analysis, 41
cross-media pollutants, 39
fats, oil, and grease, 104
techniques, 112
general pretreatment regulations,
16
local limits, 16
management strategies, 33
neutralization, 123
nutrient removal, 225
off-site, 39
organics, technologies, 192, 193
physical separation, 34
process monitoring, 40
regulations, 15
safety, 39